Front Offic Assignments

Chris Baird and Linda Carter

Pitman

PITMAN PUBLISHING
128 Long Acre, London, WC2E 9AN

© C. Baird and L. Carter 1988

First published in Great Britain 1988

British Library Cataloguing in Publication Data

Baird, Chris
    Front office assignments.
    1. Great Britain. Hotel industries.
    Receptionists. Duties
    I. Title   II. Carter, Linda
    647'.6

    ISBN 0-273-02936-3

Printed in Great Britain at The Bath Press, Avon

Contents

Acknowledgements

I would like to express my thanks to the following individuals and organisations for their help, and for their permission to reproduce information, illustrations and documentation.
Access
American Express
Dukes Hotel, Bath
Formula Two (London) Ltd
Hilary Firth HCIMA – for Chapter 9
Lloyds Bank, Walthamstow
Lyndale Hotel
Royal Garden Hotel
Starcrown Hotels
The Capital Hotel
Trust House Forte Hotels.

Key

The following terms, as commonly used in the industry, are used throughout this text.

S   Single
D   Double
T   Twin   Where any of these are followed
TR   Triple   by B the room also has a bathroom.
STE   Suite

Preface

This collection of assignments, case studies and exercises has been put together as an aid to student centred learning.

Every major examining body is moving towards assessment or testing through in-tray exercises, and it has become increasingly important for students to be able to judge a situation, and make a decision, rather than to rely upon traditional recall. It is hoped that this book will be useful both in the classroom situation, and at home, for those students who like to work at their own pace.

The assignments have been pre-tested on students taking City and Guilds 720 and BTEC Diploma. The times quoted are approximate, and will vary with the ability of the group – the time is meant as a guide rather than a criteria.

Thanks are due to Hilary Firth, Senior Lecturer in Accommodation Services at Southfields College for her contribution of Chapter 9. Hilary is an Assessor and Chief Examiner for City and Guilds 708, and a moderator for EMFEC. We are all very conscious of the need for co-operation and understanding between the front office and accommodation services department.

Many of the assignments lend themselves to further expansion, and it is hoped that both staff and students will see this potential, and enjoy the opportunity to participate in actual events.

Suggested answers are available at a nominal charge from the authors at:
    Department of Hotel Administration/Tourism and Leisure
    Waltham Forest College
    Forest Road
    London E17 4JB

1 The Hotel Industry and the Law

1 Categories of Hotels

Divide into five working groups, and allocate each group a hotel belonging to one of the following:
- an international hotel chain
- a national hotel company
- a small hotel group
- an independent hotel
- a consortium

Tasks

1 Describe what is meant by each of the different categories of hotels.

2 List the advantages of the type of establishment allocated to your group.

3 Name one hotel in each category.

4 One person in each group is to write and one person is to telephone the hotel for a brochure.

Write a report on the telephone call, commenting on the
- speed
- efficiency
- product knowledge
- sales technique

of the person who received the call at the hotel.

5 When you receive the brochure:

(a) Write a report commenting on:
- the envelope
- the brochure (what sector of the market do you think it is aimed at?)
- the tariff
- the letter
- the length of time taken since it was written/telephoned for
- any additional information sent.

(b) Find out the AA star rating of the hotel. List the minimum requirements of this category.

(c) Nominate one person to give a verbal presentation to the rest of the class on your findings and allow time for questions.

(Estimated time for assignment: 3–5 weeks)

2 Hotel and catering outlet

From the following list, each pick two categories and name an establishment in your area with which you are familiar. Try to cover all categories between you.

- traditional restaurant
- fast food restaurant
- industrial caterer
- hospital/welfare caterer
- transport caterer
- public house
- wine bar
- cocktail bar
- traditional club
- night club
- conference centre
- leisure centre
- educational establishment
- residential establishment

1 For each category nominate one person to write or telephone and find out about the facilities available, and fees if appropriate, for each establishment chosen.

2 As a group prepare a job opportunities folder. Check local papers for staff advertisements which could possibly be filled by a successful BTEC or City and Guilds 720 student.

(Estimated time for assignment 3–4 weeks)

3   Career paths

Tasks

1 Draw a diagram or flow chart showing a possible career path for a college leaver with City & Guilds 720 or BTEC Diploma, in:
  (a) small hotels
  (b) a large hotel chain

40 minutes

2 Look in the trade press, professional magazines, and local and national papers and compile a file of vacancies suitable for each stage of the career path.

3–4 weeks

3 List the major professional and technical associations which could be helpful to a student aiming for the top in the hotel and catering trade.

1 hour

4 Match these qualifications with the most appropriate vacancies:

IPM   housekeeper
HCIMA   commis waiter
C & G 707/1   hotel manager
BTEC Cert   commis chef
C & G 708   personnel manager
RIPHH   trainee front office clerk

10 minutes

5 Describe how each of the following organisations help to maintain professional standards:
  1 HCIMA
  2 ABTA
  3 HCTB
  4 CGLI

1 hour (including research)

4 Front office organisation

1 Using hotels A, B and C in Appendix 1, draw an organisation structure chart for the front office area of each, including, where appropriate, the following sections:
- management
- reception
- billing office
- cashier
- advance reservations
- telephone switchboard
- hall porters, luggage porters, linkman, pages, life attendant, night porter, cloakroom attendant,
- enquiries office
- night auditor
- guest services

2 On your return from industrial release compare the hotel where you worked with the most similar organisation chart, making a note of any differences.

45 mins–1 hour

5 Duty Rotas

You are the supervisor in a department employing eight staff. The rota, which it is your job to compile, can be designed to work for a three week re-occurring period for ease of operation. Each member of staff should have one week-end (Saturday and Sunday) off in three, and three other days off over the period – ideally this should be one day one week and two the next. Staff should always work early before their day off and late after their day off. No one should work more than nine days without time off, and no split duties may be worked, although the occasional middle shift is permissible, especially at weekends when it is not very busy. There should be, as near as possible, equal numbers of staff on each shift, or slightly more on early than on late – early shift being the busiest time. You should avoid, where possible, more than two earlies or two lates in a row.

Task

Construct a rota based on these guidelines. You may wish to use the suggested layout below, or construct your own.
Suggested layout

	Mon	Tues	Wed	Thur	Fri	Sat	Sun	_Week 2_	_Week 3_
1									
2									
3									
4									
5									
6									
7									
8									

1 hour

6 Complaint handling

You are on duty as the Receptionist at the Bleyford Hotel. A very angry lady stamps into the lobby, and bangs her umbrella down hard on the desk, commanding attention.

She immediately launches into a furious tirade directed at the inefficiency of the staff, the poor management of the hotel and the inconvenience to which she has been subjected.

It transpires that the evening staff last night had booked her into a nearby hotel because they were overbooked. She did not receive any telephone calls, although she knows several people have called her. Her luggage was not sent on as promised and the other hotel omitted to wake her up this morning.

To add insult to injury her friend arrived at your hotel *after* her and was given a room.

Tasks

Divide into pairs or groups and discuss the following.
1 Describe how you would deal with this situation – what would you say to her?
2 What suggestions would you make to ensure it does not happen again?
3 What social skills would need to be employed in this situation?
4 What would you do to try to ensure the guest booked with you again?

1 hour

7 Personal presentation

You are 18 years old and have just completed a course at college where you successfully gained your 720 Diploma in Hotel Reception and Front Office Procedures.

You have been very lucky in securing a post with a well-known hotel consortium, your own particular hotel being a 120 bedroom privately-owned establishment. The work is interesting and quite varied, and so far you have not encountered anything which you cannot handle. The staff all seem very friendly and apart from the shift work, which you are finding a bit tiring, the job does not seem too hard. You know you were lucky to find the job, and are quite pleased with it. Your first few weeks were spent in the back offices, familiarising yourself with the advance reservations system and the switchboard, but this week you have been on the desk which has been more interesting. However, the Head Receptionist has asked to see you. She makes a few enquiries about your general progress and then launches into an attack on your appearance. She says that your clothes are unsuitable for the job you are doing, and that your hair must be cut and tidied up – she implies that you do not know how to dress (which hurts you since you spend a lot of money on clothes, and all your friends rate you a 'snappy dresser'). You spend a lot of time on your hair, wash it every other day, and it is always cut in the latest fashion. You are quite stunned by this attack, and reply that if they hadn't liked the way you looked they needn't have taken you on. The Head Receptionist is 'past it' anyway, in your opinion, and probably thinks you should be all be wearing full length skirts. You are off for the next two days and she says that she is looking to see some improvement when you return.

You can see nothing wrong with the way you look, and you return to work after two days having made no alterations to your appearance. She asks to see you at 15.00 hours when you finish.

Tasks

1 What do you intend to say? As a group discuss how you think you should handle the situation.
2 Put yourself in the position of the Head Receptionist. What do you think she is thinking and what picture does she have of the student. Discuss within your groups.

1 hour

8 Theft

Mr and Mrs Bishop are staying in your hotel for seven days and have taken advantage of the safe deposit facilities which are offered by the hotel. They arrive back late after an evening at the theatre, followed by supper, and find that the night auditor is busy checking in a large group of Italian students. Mrs Bishop wants to deposit her jewellery and asks for the master key to the safe deposit box. The auditor says he will telephone them in their room as soon as he can leave the desk. Mr and Mrs Bishop go upstairs, and after half an hour, having heard nothing, they ring down to the front desk but there is no reply. Mr Bishop goes down and is told by the night porter that the auditor has gone for a break. After several more attempts to contact the night auditor they decide to go to bed and return the jewellery to the box in the morning.

When they wake up in the morning the jewellery has gone, along with their travellers' cheques and small change.

Tasks

1 Explain the legal implications of this situation.
2 How would the situation differ if the establishment was a private hotel?
3 As the auditor involved in this situation write two reports:
   (a) one to management – explaining how and why this situation has occurred;
   (b) one for the police who are investigating the theft.

1 hour

9 Hotel tariffs

The hotel where you are employed displays the following tariff by the front door:

Single Room	£20–£25 inc VAT
Twin/Double	£35–£50 inc VAT
Extra Bed	£7.00 inc VAT
Table D'Hote Lunch	£5.50
Table D'Hote Dinner	£7.50

During the week of the County Show the management has decided to let rooms on the basis of half board or full board only, and to impose a minimum booking period of three days.

Tasks

1 Re-design the tariff in order to comply with the law.
2 State the Act which governs the innkeeper's obligation to show the tariff.
On the first night of the County Show the following accommodation was let:

12 S @ Demi Pension
10 S @ Full Pension

17 T/D @ DP
24 T/D @ FP
2 TR @ DP

3 Calculate the revenue from accommodation for that night.
4 Extract the VAT figure for the same night.

1 hour

10 Contracts of booking

The hotel where you work is very busy and all unconfirmed reservations carry a 6.00pm release.

You are on late shift and at 17.45 four ladies arrive (Ms Pam, Ms Connolly, Ms Messum and Ms Marshall) asking if you have accommodation and dinner for tonight. According to the arrival list there are three rooms with a 6.00pm release:

Carveth   TB   106
Faulkner   TasS   410
Pound   DB   321

You decide to take a chance in order to maximise occupancy, and release 106 and 410. The ladies register and go upstairs.

Tasks

1 List *exactly* the procedure you would follow to cancel the original bookings and sustitute the new one.
2 Mr and Mrs Carveth arrive at 17.55
    (a) What should you do?
    (b) What are the legal implications?
At 18.15 Mr. Faulkner arrives. He insists he was not told about the 6.00pm release policy or he could easily have checked in earlier. He has told his business associates he is staying with you, and his company is sending a car for him tomorrow morning. He is very angry and talking about suing the hotel. You have already let Room 321 and there is no other unconfirmed room available.
3 Explain how you will deal with this situation.
4 What are the legal implications?

1 hour

2   Reservations

1   Reservation forms

Shown on pp. 13–14 are three different types of reservation form.

Tasks

1 State what type and size of hotel you imagine each form is for.
2 State what system you think will be employed in Advance Reservations in each case, giving reason.

30 minutes

RESERVATIONS LOADING CARD

A

Tick Box if Reservation is for Group or Tour ☐

PROMOTION NO.
SIL ACCOUNT NO.
COMMISSION AREA

ROOMS

IN DATE

OUT DATE

RESERVATION NO. R

TOUR NO. T

BOOKED BY

MARKET SEGMENT   AUTHORITY (IF ID)

RATES

TOUR NAME (if applicable)

PAYMENT TYPE   If Part Account to Company or Voucher, give brief details of charges acceptable to Company.

For Account to Company/Voucher Payment Types (FC, AC, FV, AV, QC, QV) only

A/C NAME

A/C ADD

For Travel Agent Bookings only. If same as Account Name & Address write A/C

T/A NAME

T/A ADD

COMMISSION RATE   OVER-RIDE COMMISSION (Y/N) ☐

REPLY NAME

REPLY ADD

BOOKING METHOD ☐   RELEASE TIME   SOURCE CODE

INITIALS

(PLEASE ENTER GUEST DETAILS ON REVERSE)

13

C

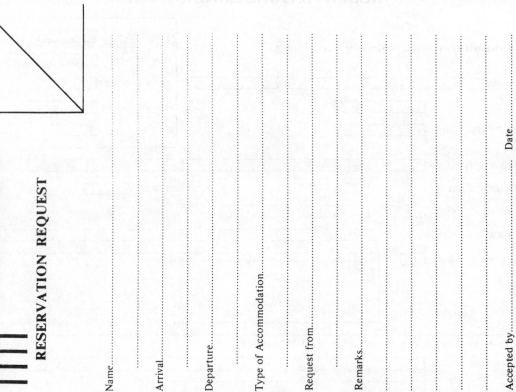

CAPITAL HOTEL

RESERVATION REQUEST

Name...

Arrival...

Departure...

Type of Accommodation...

Request from...

Remarks...

....

....

Accepted by... Date...................

B

Lyndale Hotel

AND RESTAURANT

RESERVATION FORM

Name ....

Address ....

....

....

Tel. (day) .... (evening) ....

Please reserve the following number of rooms:

.... Single .... Twin

.... Double .... Family

Number of persons:

.... Adults .... Children

Number of nights ....

Dates from .... to ....

Additional requirements (cots, special diets etc.): ....

I enclose £.......... being a deposit of £25 per person (children exempt) towards the cost of the holiday.

Signed ....

Date ....

Your reservation constitutes a legal contract which requires you to pay for the reserved accommodation in the event of cancellation unless the accommodation can be re-let.

We have arranged a special holiday cancellation insurance scheme for a very small premium, details will be sent with your confirmation.

2 Reservation recording

Details of an expected client taken by phone:

Ms M Bate
SB Arr. 9/9
Five nights
Tel: 01 521 2134
Paying own account by American Express
Requested a non-smoking room
1700 Arrival

Tasks

1 Complete the three reservation forms, A, B, and C from the previous assignment, using the information above.
2 Why do reservation forms differ from one establishment to another?

20 minutes

3 Methods of confirmation

Some reservation records are shown below.

1 Draft a suitable reply to be sent by telex to Mr Todd (B) to confirm his reservation

2 Complete the standard confirmation form (A) to Mr and Mrs Granger (C) showing reservation details as necessary.

3 Write a suitable reply to be sent to Mr and Mrs Holmes (D) confirming their booking.

4 Mr Fisher's (E) confirmation is to be sent by telemessage. Write the message as it would appear.

1 hour

NORFOLK TOWERS HOTEL

34 NORFOLK PLACE LONDON W2 1QW
Telephone: 01-262 3123 Telex: 268583 NORTOW

CONFIRMATION OF RESERVATION

Your Ref: . . . .

Our Ref: . . . .

Date . . . .

We have pleasure in confirming/offering you the following reservations:

Date of Arrival	No. of Nights	Name/s	Room	Daily Rate	Remarks

DAILY RATE: Inclusive of Continental Breakfast and Service Charge but exclusive of Value Added Tax.
RESERVATIONS: Accommodation will be automatically released at 15.00 hours unless a deposit has been received or arrival is guaranteed by Company.
DEPOSITS: The first night's charges are required in advance as a holding deposit along with your letter of confirmation.
ON ARRIVAL: Rooms may not be available before mid-day.
ON DEPARTURE: Rooms should be vacated by mid-day.
PERSONAL CHEQUES: We regret that personal cheques can be accepted only if prior arrangements have been made or on production of your Banker's Cheque Card provided your account is within the limits set by the card.
We regret it has become necessary to pay for all accommodation in advance on arrival. We apologise for any inconvenience.
We reserve the right to provide alternative accommodation and services of higher or similar standard. Rates shown on this confirmation are subject to alteration, without prior notice, only when Government Taxes are changed.
Should you require further information regarding this reservation please contact the hotel directly.
Assuring you of our best attention at all times.

ROOM CODE:
S : Single
SB : Single/Bath
SS : Single/Shower
T : Twin
TB : Twin/Bath
TS : Twin/Shower
D : Double
DB : Double/Bath
DS : Double/Shower
TR : Triple
TRB : Triple/Bath
TRS : Triple/Shower
S/occ : Single Occupancy

A

B

RESERVATION RECORD

NAME (Block Letters) Mr. P Todd

RESERVED BY STEELY Co. - by telex

ADDRESS .... TEL. ....

REQUIREMENTS SB .... No. OF NIGHTS 8

ARRIVAL DATE 1/4 .... No. OF NIGHTS 2

ARRIVAL DATE 30/4

RATE £85.00 + VAT

REMARKS Return guest
a/c to co.

DATE 1/3 SIGNATURE [signature]

DO YOU NEED ANYTHING ELSE?

☐ CHARTED ☐ CHECKED

C

RESERVATION RECORD

NAME (Block Letters) M/M GRANGER

RESERVED BY

ADDRESS 19 Parkway
Wellborough, Hants TEL 821471

REQUIREMENTS DB

ARRIVAL DATE 2/4 .... No. OF NIGHTS 5

RATE £85.00 + VAT

REMARKS Anniversary
Deposit Paid £50.00

DATE 1/3 SIGNATURE [signature]

DO YOU NEED ANYTHING ELSE?

☐ CHARTED ☐ CHECKED

D

RESERVATION RECORD

NAME (Block Letters) Mr/s Holmes

RESERVED BY Whizco Ltd

ADDRESS Ferry Lane Trading Estate
Brigthorpe Lancs TEL 413214

REQUIREMENTS TB

ARRIVAL DATE 8/4 .... No. OF NIGHTS 1

ARRIVAL DATE .... No. OF NIGHTS

RATE £85.00 + VAT

REMARKS Will meet Banqueting Manager
on arrival Wife is disabled
- (Room near lift)

DATE 20/3 SIGNATURE [signature]

DO YOU NEED ANYTHING ELSE?

☐ CHARTED ☐ CHECKED

E

RESERVATION RECORD

NAME (Block Letters) MR FISHER

RESERVED BY Mr Fisher - by phone

ADDRESS (Travelling - at Wells Park
Hotel until 1/4) TEL 91241

REQUIREMENTS SB

ARRIVAL DATE 15/4 .... No. OF NIGHTS 3

ARRIVAL DATE .... No. OF NIGHTS

RATE £85.00 + VAT

REMARKS Will pay in advance
on arrival at 1pm

DATE 15/3 SIGNATURE [signature]

DO YOU NEED ANYTHING ELSE?

☐ CHARTED ☐ CHECKED

4 Stop-go chart

The computer is out of action and the stop-go chart is the only means of reference. Examine this section of the chart.

	1st	2nd	3rd	4th	5th	6th	7th
S						*	*
T		*	**		*		
D				**	**	*	*
STE				**	*		

  * Take 5 only
** Full. See Manager if VIP

Task

1 On the basis of this section of the stop-go chart, state how you would deal with the following requests:
- D   1st – 3 nights
- S   3rd – 5 nights
- S   2nd– 7 nights
- D   5th – 3 nights
- STE 1st – 4 nights
- 4S   6th – 2 nights

2 Explain why you have taken these decisions.

3 Give three advantages and three disadvantages of using a stop-go chart.

20 minutes

5 Reservation and telex reply

A message left for the Reservations Office by the night porter read as follows:

Mr P Wilmott wants a twin room with bath for the 12th Sept for 3 nights. (I told him the price was £95 + VAT). I showed him 313 and he liked it. He will arrive at 10pm and said that his company will pay for his room and breakfast. I said you will contact to confirm etc.

TLX 2311562 Company Name: Challis & Murphy (Building Contractors)
From John
Night Porter
Time 2345 1/9

Tasks

1 Complete the reservation form from the above information.
2 Draft a suitable reply to be sent by telex.

30 minutes

ROOM RESERVATION

NAME..	ARRIVAL..
ADDRESS..	DEPARTURE....................................
CITY..	PHONE...
SINGLE ☐ TWIN ☐	NO. OF PERSONS...........................
DOUBLE BED ☐ SUITE ☐	RATE...
REMARKS..	
RESERVATION REQUESTED BY...........................	
COMPANY...	PHONE...
ADDRESS..	

BILL TO..

	RESERVATION TAKEN BY.....................................
PHONE ☐ VERBALLY ☐ DATE........................	
TO BE CONFIRMED BY HOTEL ☐ BY GUEST ☐ NOT ☐	

6 Accommodation availability

On the 1st September you have two TB and two SB available. The following requests for accommodation are received.

- Mr and Mrs A Devine   TB   7 nights   request from TransAmerica Travel
- Mssrs Glasgow and Arkless   TB   4 nights   from Avitron @ corporate rate
- Ms Seward   TB   3 nights   Freesale from HBT   Allocations
- Lord and Lady Plummer   TB   5 nights
- Senor J Latorre   TB   4 nights   for his annual visit
- Mrs A Horner   SB   4 nights   Secretary to Sales Director of Techno Co.
- Mr G Lund   SB   1 night   attending annual Police Ball
- Mr G Herring   SB   4 nights   Deasons Distillery
- Sir Wilfred Barr   SB   2 nights   always stays if in town
- Miss I Hanrahan   SB   1 night   has late appointment with Banqueting Manager

Tasks

1 Checking through the Guest History Cards you see Mr Herring is on the 'black list' because he left with the hotel towels. Write a letter to him saying you cannot accept his booking.

2 To whom will you offer the two TB and two SB? Give reasons for your choice.

3 In what order will you waitlist the remaining clients? Give a full explanation of your eventual list.

1 hour

7 Bedroom book

- Brought forward from September:

Messrs Roebuck, Ash and Leather   3SB   leaving on 2nd October
Mr/s Still   TB   leaving on 3rd
Miss Ward   S   leaving on 2nd
Mr/s and Misses Ellis   DB and 2S   leaving on 3rd

- Arrivals : bookings: from 1st October

Mr Curtis   S   2 nights
Mr/s Hicks   DB   1 night
Messrs Buck, Baillie and Cross   3SB   3 nights
Mr/s Halford   TB   1 night

- Arrivals: bookings from 2nd October

Miss Lemar   S   1 night
Mr/s Herbert   T   2 nights
Mr/s Hughes   T   1 night
Misses Alexander   TB   2 nights
Messrs Roper and Smith   2S   1 night
Mr/s Dale   DB   2 nights
Rev/Mrs Binns   TB   2 nights
Dr/Mrs Simon   D   2 nights

- Arrivals: bookings from 3rd October

Messrs Bond, Cape and Swan   3SB
Mr/s Allen   DB
Mr Blake   SB
Miss Saul   S
Mr/s Deal   T

- Enquiry from See-All Tours: can you do 4S, 3TB, 1D, 2T for three nights from 1st October? If not, have you an alternative for them?

Task

Enter the booking in the diary/bedroom book provided (p. 22). If you are unable to accommodate clients with their first choice provide an alternative and state why you chose it.

45 minutes

Bedroom Book

1/10				2/10				3/10			
Room	Type	Names	Stay	Room	Type	Names	Stay	Room	Type	Names	Stay
101	TB			101	TB			101	TB		
102	DB			102	DB			102	DB		
103	S			103	S			103	S		
104	T			104	T			104	T		
105	S			105	S			105	S		
106	TR			106	TR			106	TR		
107	TB			107	TB			107	TB		
108	TRB			108	TRB			108	TRB		
109	SB			109	SB			109	SB		
110	SB			110	SB			110	SB		
111	TB			111	TB			111	TB		
112	SB			112	SB			112	SB		
114	D			114	D			114	D		
115	S			115	S			115	S		
116	DB			116	DB			116	DB		
117	S			117	S			117	S		
118	SB			118	SB			118	SB		
119	S			119	S			119	S		
120	TR			120	TR			120	TR		
121	SB			121	SB			121	SB		
122	S			122	S			122	S		
123	SB			123	SB			123	SB		
124	S			124	S			124	S		
125	T			125	T			125	T		
126	D			126	D			126	D		

8 Traditional/conventional chart

The hotel where you work operates a conventional chart booking system.

Bookings

TB	1st–4th inc	Jones
DB	6th–3 nights	Williams
SB	3rd–6th	Green
D	2nd–6th	James
D	8th–11th inc	Reed
D	6th–8th	Brown
D	5th–9th	Grey
T	2nd–4th inc	Carter
DB	3rd–6th	Hawkins
T	6th–8th	Howe
TRB	2nd–4 nights	Day
D	6th–9th	Plummer
TB	1st–5th	Small
T	3rd–6th	Croucher
TR	4th–7th inc	Thatcher
T	8th–13th	Jackson
S	3rd–6th	Smith
T	5th–11th	Baird
S	7th–11th	Dix
TB	5th–11th inc	Barker
S	9th–10th	Smith
STE	6th–12th inc	Wilson
S	5th–2 nights	Parker
S	7th–10th	Bland
S	1st–5th	Dickson
DB	3rd–5 nights	White
S	7th–9th	Seldon

TB 5th–7th   Sutherland

D   3rd–6th   Glover

S   3rd–7th   Hooper

TR 5th–10th   Davies

TB 3rd–6th   Cooper

TB 5th–7th   Cox

Task

1 Enter the bookings above onto the chart provided.
2 List the advantages of a chart of this type.
3 List the disadvantages of the chart.
4 If you are unable to accommodate all the clients in the rooms which they booked make alternative arrangements, stating what you have done, and why.

45 minutes

	Mon	Tues	Weds	Thur	Fri	Sat	Sun	Mon	Tues	Weds	Thur	Fri	Sat
101 TB													
102 STE													
103 T													
104 S													
105 D													
106 DB													
107 TR													
108 SB													
109 TRB													
110 S													
111 D													
112 TB													
113 STE													
114 T													
115 S													
116 D													
117 DB													
118 TR													
119 TB													

9 Density chart

- Arrival 1st
Jones M/M P	TB	4
White MR J	SB	5
Green M/M T	TRB	2
Brown MR R	SB	6
Read M/M R	2DB	4
Smith M/M D	DB	2
Abel M/M S	TB	1

- Arrival 2nd
Apple MS P	SB	4
Humphreys MR S	SB	6
Fernandez M/M H	TB	5
Goddard M/M C	DB	4
Wills M/M M	TRB	2
Baird M/M P	TRB	4
Carter M/M I	DB	6
Bayram M/M B	TB	7

- Arrival 3rd
Kellogs LTD	4TB	5
Barker MR T	SB	4
Laing M/M J	TRB	9
Williams M/M P	DB	6
Hawkins M/M C	TB	5
Seldon MS D	TB	7
Stickles MS D	SB	2

- Arrival 4th
Jones M/M R	2TRB	7
Bland M/M F	DB	4
Herring M/M R	TB	7
Spikings M/M T	DB	6
Yuffs MR R	SB	5
Matthews MR T	SB	8
Middleton M/M W	TB	6

- Arrival 5th
Milne M/M J	3DB	2
Kirman M/M S	TB	7
Gray MR K	SB	6
Hill MR S	SB	5
Welch M/M R	TB	7

Cole M/M R	DB	4
Dent M/M J	TB	3
Cowling M/M R	TRB	4

- Arrival 6th
Palmer MR D	SB	2
Jarvis M/M J	TB	1
Barley M/M K	TRB	7
Simpson MR P	SB	2
Acey Mr J	SB	2
Winter M/M P	TB	5
Burkett M/MP	SB	6
Sopp M/M M	DB	5

- M/M Fernandez extend their booking to seven nights from the 2nd to the 9th.
- Mr J White has cancelled his stay.
- Mr J Acey is now arriving on the 4th for 6 nights.
- Kelloggs now require a SB for the same period as the rest of their group.

Tasks

1 Enter the bookings on density chart A – For TRB use Semi Suites.

A

SUNDAY	MONDAY	TUESDAY	WEDNESDAY	THURSDAY	FRIDAY	SATURDAY
Twins (55)	Twins (55)	Twins (55)	Twins (55)	Twins (55)	Twins (55)	Twins (55)
Doubles (23)	Doubles (23)	Doubles (23)	Doubles (23)	Doubles (23)	Doubles (23)	Doubles (23)
Singles (23)	Singles (23)	Singles (23)	Singles (23)	Singles (23)	Singles (23)	Singles (23)
Suites Semi (7)	Suites Semi (7)	Suites Semi (7)	Suites Semi (7)	Suites Semi (7)	Suites Semi (7)	Suites Semi (7)
Suites (3)	Suites (3)	Suites (3)	Suites (3)	Suites (3)	Suites (3)	Suites (3)
28	28	28	28	28	28	28
GROUPS	GROUPS	GROUPS	GROUPS	GROUPS	GROUPS	GROUPS

45 minutes–1 hour

SB	1/11–4/11	Wilmott
TB	1/11–3/11	Jones
TB	1/11–5/11	Green
SB	1/11–2/11	Brown
TRB	1/11–3/11	Falco
SB	1/11–6/11	Johnson
DB	2/11–6/11	Bayshaw
TB	2/11–7/11	Kerridge
TB	2/11–3/11	Dow
TRB	2/11–6/11	Keys
SB	3/11–9/11	Salaria
DB	3/11–8/11	Atheris
TRB	3/11–7/11	Delort
TB	3/11–8/11	Springham
SB	4/11–10/11	Mason
DB	4/11–9/11	Obertelli
TRB	4/11–8/11	Pajares
TB	5/11–10/11	Hazelby
SB	5/11–12/11	Fearnley
DB	5/11–11/11	Manniong
TB	5/11–6/11	Humphreys
5TB	6/11–9/11	
6SB	6/11–9/11	Texaco
2DB	6/11–9/11	
TB	6/11–11/11	Fernandez
TRB	6/11–7/11	Ashley
TB	7/11–12/11	Evans
DB	7/11–8/11	Burnett
8SB	7/11–8/11	Van Noorden
TB	8/11–11/11	Christianson
TRB	8/11–10/11	Vanessi
TB	8/11–18/11	Thunhurst
DB	8/11–12/11	Wells
SB	9/11–16/11	Atkinson
3SB	9/11–13/11	Alexander
2TB	10/11–14/11	Pendle
DB	10/11–13/11	Hughs
SB	11/11–15/11	Beecham
TB	13/11–18/11	Williams
SB	13/11–18/11	Plummer
DB	13/11–19/11	Prewitt
TB	14/11–17/11	John
SB	14/11–19/11	Read
5TB	15/11–17/11	
6SB	15/11–17/11	Koch Oil
SB	15/11–20/11	Powell

2DB   15/11–18/11 King
TRB   17/11–19/11 Jackson
TB   17/11–18/11 Jensen
SB   17/11–20/11 Peel
DB   17/11–20/11 Saville

- Texaco require another two twins for the same period as their group.
- Cancel Mr Atkinson's reservation.
- Mr Beecham requires another single for the same period.
- Mr Dow is arriving on 2/11 and now wants to extend by three nights.
- Ms Hazelby, arriving on the 5th, now wants to arrive on the 4th.

2 Enter the bookings on density chart B. For TRB use semi-suites.

B

SUNDAY	MONDAY	TUESDAY	WEDNESDAY	THURSDAY	FRIDAY	SATURDAY
Twins (55)	Twins (55)	Twins (55)	Twins (55)	Twins (55)	Twins (55)	Twins (55)
Doubles (23)	Doubles (23)	Doubles (23)	Doubles (23)	Doubles (23)	Doubles (23)	Doubles (23)
Singles (23)	Singles (23)	Singles (23)	Singles (23)	Singles (23)	Singles (23)	Singles (23)
Suites Semi (7)	Suites Semi (7)	Suites Semi (7)	Suites Semi (7)	Suites Semi (7)	Suites Semi (7)	Suites Semi (7)
Suites (3)	Suites (3)	Suites (3)	Suites (3)	Suites (3)	Suites (3)	Suites (3)
28	28	28	28	28	28	28
GROUPS	GROUPS	GROUPS	GROUPS	GROUPS	GROUPS	GROUPS

1 hour

10   Advance deposits

You have just been appointed Head Receptionist at the Chrislin Hotel. When you arrive on 1st January you find that the system of dealing with deposits for advance bookings is as follows:
- enter the amount paid in the remarks column of the Diary
- place an envelope marked with the intended visitor's name, containing the deposit in the safe.

The sixteen envelopes which were in the safe agreed with the notes made in the Diary. Twelve contained cash and four contained cheques, (three cheques were dated between the 20th and 30th September and one was dated for next February 18th).

The retiring Head Receptionist told you that the deposit was always handed back to the visitor on arrival, by her personally, and nothing had ever gone wrong with the system.

Tasks

1 Explain in detail, using diagrams if necessary, what system you would substitute.
2 Explain why you would not accept the current procedure.

45 minutes

11 Diary and density chart

Examine the following information.

Tasks

1 Enter the reservations onto the density chart and diary.
2 During the day you receive the following cancellations and alterations.
Make the necessary adjustments to the records.
- Bailet, Arigo and Bellaton are unable to get a flight and will arrive on the 2nd for six nights.
- Ms Garrard, cancelled due to business plans.
- Mr Lee telephones and asks for flowers and champagne in room for his anniversary.
- Mr Holmes telephones to guarantee his reservation by American Express and gives his card number.
- Mrs Brittain wishes to bring a friend and requires a Twin B (which is available).

3 In what type of hotel (approximate size and star rating) would you expect to see a diary in operation?

1 hour

1ST MARCH NAME	Type of Room	No. of Nights	RATE	How Booked	Confirm-ation	Room No.	REMARKS
BELL M/M	DB	1	£55	T16/12	C/F		Close to Mr Toynton
LEE M/M F.	Suite	7	£85	T16/12	Tx 10/2		
BRITTAIN MRS B.	SB	5	£40	T16/12	C/F		
TOYNTON MR B.	SB	1	£40	T4/1	C/F		Close to M/M BELL
BAILET MR S.	SB	6	£40	⎫	⎫		a/c to Reed Builders
ARIGO MR M.	SB	6	"	⎬ 26/1	⎬ C/F		" " "
BELLATON MR P.	SB	6	"	⎭	⎭		" " "
BAYRAM MR K & FAM.	2TB	3	£55	L 28/1			
TODD M/M P.	TB	8	£55	TLY 28/1			
GARRARD Ms A.	SB	4	£40	Tel 31/1			Wants high floor
HUTCHINSON MRS C.	SB	1	£40	Tel 1/2			
WATKINS M/M G.T.	S.Suite	2	£65	✓ 10/2			Deposit £50
KATHERENS M/M W.	S.Suite	5	"	Tel 18/2			
DAWSON M/M K.	TB	2	£55	T/x 26/2			Liked #43
HOLMES MR A.	DB	1	"	T. 28/2			Single Occ.

<table>
<tr><td>

RESERVATION RECORD

NAME: D. Johnson
(Block Letters)

RESERVED BY: D. Johnston

ADDRESS: The Priory – Welton, Humberside. TEL. 20974

REQUIREMENTS: SB

ARRIVAL DATE: 1/3 No. OF NIGHTS

ARRIVAL DATE: 5/3 No. OF NIGHTS 4

RATE: £40

REMARKS: Arrival 2pm. Own A/c.

DATE: 28/2 SIGNATURE: Rb

DO YOU NEED ANYTHING ELSE?

☐ CHARTED ☐ CHECKED

</td><td>

RESERVATION RECORD

NAME: M/M. J. Stratton
(Block Letters)

RESERVED BY: Hutchco Ltd

ADDRESS: 149 the Parkway, Bolton, TEL. 25971

REQUIREMENTS: TB

ARRIVAL DATE: 1/3 No. OF NIGHTS

ARRIVAL DATE: 6/3 No. OF NIGHTS 5

RATE: £40 *

REMARKS: Single Occupancy
* Regular Guest always has discounted rate.

DATE: 28/2 SIGNATURE: Rb

DO YOU NEED ANYTHING ELSE?

☐ CHARTED ☐ CHECKED

</td></tr>
<tr><td>

RESERVATION RECORD

NAME: MR. G. Leeson
(Block Letters)

RESERVED BY: Helen of HBT.

ADDRESS: 20 High Street, Luxton TEL. 991271

REQUIREMENTS: TB

ARRIVAL DATE: 1/3 No. OF NIGHTS

ARRIVAL DATE: 2/3 No. OF NIGHTS 1

RATE: £55

REMARKS: GTD per Agent for late arr.

DATE: 28/2 SIGNATURE: Rb

DO YOU NEED ANYTHING ELSE?

☐ CHARTED ☐ CHECKED

</td><td>

RESERVATION RECORD

NAME: Ms Jung Wallis
(Block Letters)

RESERVED BY: Clothco

ADDRESS: Mill Lane Trading Estate Mill Road, Elland. TEL. 0269 65754

REQUIREMENTS: SB

ARRIVAL DATE: 1/3 No. OF NIGHTS

ARRIVAL DATE: 8/3 No. OF NIGHTS 7

RATE: £40

REMARKS: Wants a Room at the front of hotel

DATE: 28/2 SIGNATURE: Rb

DO YOU NEED ANYTHING ELSE?

☐ CHARTED ☐ CHECKED

</td></tr>
</table>

FEB/MARCH

28th SUNDAY	1st MONDAY	2nd TUESDAY	3rd WEDNESDAY	4th THURSDAY	5th FRIDAY	6th SATURDAY
Twins (55)	Twins (55)	Twins (55)	Twins (55)	Twins (55)	Twins (55)	Twins (55)
Doubles (23)	Doubles (23)	Doubles (23)	Doubles (23)	Doubles (23)	Doubles (23)	Doubles (23)
Singles (23)	Singles (23)	Singles (23)	Singles (23)	Singles (23)	Singles (23)	Singles (23)
Suites Semi (7)	Suites Semi (7)	Suites Semi (7)	Suites Semi (7)	Suites Semi (7)	Suites Semi (7)	Suites Semi (7)
Suites (3)	Suites (3)	Suites (3)	Suites (3)	Suites (3)	Suites (3)	Suites (3)
28	28	28	28	28	28	28
GROUPS	GROUPS	GROUPS	GROUPS	GROUPS	GROUPS	GROUPS

HOWDEN PARK HOTEL

Internal
Memorandum

To Reservations   Your ref   Date 28/2
                                     (10 pm)

From K Williams   My ref
       Gen Manager

Subject Mega Tours

Please reserve   18 Twins
                    8 Singles
                    4 doubles

for the 1st march for 4 nights - booking
from Central Reservations office, rates to
follow, phone Carol at 9 am in the
morning, group will arrive at approx
4 pm.

K William.

3 Check-in and Guest in House

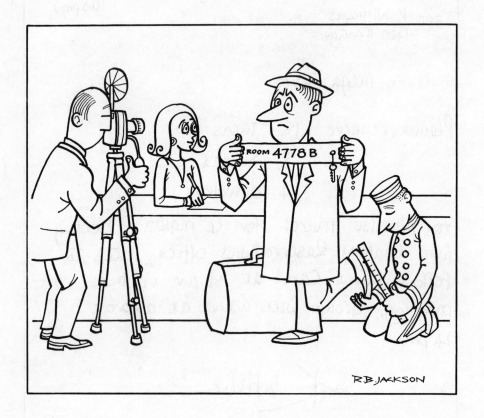

R.B. JACKSON

1 Departmental communication

While you are on duty in reception in Hotel B (see Appendix, p. 168) the following situations occur.

- Four early departures (guests leaving earlier than planned)
- One move
- Two guests extending their stay
- baby cot required for arrival tonight
- VIP client
- day let room needed again this evening
- complaint from client in old wing – always stays in new wing
- your regular aircrew contract arriving 11.00 am instead of 4.00 pm

Tasks

1 List the departments you would notify of these events.
2 List the records that would be changed.
3 List the documents which would have to be circulated.

30 minutes

2 Registration

The registration cards shown here were accepted by a trainee in your absence.

Card 1 (top left):

Length of stay	2N
Number in party	1

CASH ☑
CHEQUE ☐
A/C ☐
CREDIT CARD ☐
VOUCHER ☐

EROS HOTEL
65-73 Shaftesbury Avenue
London W1V 8EX
Telephone: 01-734 8781
Telex: 892676 VICGDN

MR. MRS. MISS: WILLIAMS John
Please Print in Full

Address: 2 The Cottage Wilmot Street
Aberdeen

City

Country _____ Nationality: Brit Car Registration Number _____
Passport No. (Aliens only) _____ Where Issued _____
Next Destination (Give address) _____

ACCOUNT CUSTOMERS ONLY – If the Company for whatever reason fails to honour this account, I undertake to be held personally liable for the full payment thereof.

Receptionist: SW. Signature: J. Williams.

Card 2 (top right):

Length of stay	1N
Number in party	2

CASH ☑
CHEQUE ☐
A/C ☐
CREDIT CARD ☐
VOUCHER ☐

EROS HOTEL
65-73 Shaftesbury Avenue
London W1V 8EX
Telephone: 01-734 8781
Telex: 892676 VICGDN

MR. MRS. MISS: Joseph -Blake.
Please Print in Full

Address: P O Box 164
Doha Qatar

City

Country _____ Nationality _____ Car Registration Number _____
Passport No. (Aliens only) 20978 Where Issued London
Next Destination (Give address) Qatar

ACCOUNT CUSTOMERS ONLY – If the Company for whatever reason fails to honour this account, I undertake to be held personally liable for the full payment thereof.

Receptionist: SW. Signature: Blake Joseph

Card 3 (bottom left):

Length of stay	7N
Number in party	1

CASH ☐
CHEQUE ☐
A/C ☐
CREDIT CARD ☑
VOUCHER ☐

EROS HOTEL
65-73 Shaftesbury Avenue
London W1V 8EX
Telephone: 01-734 8781
Telex: 892676 VICGDN

MR. MRS. MISS: Dr. P. MARTINO. (Pueblo)
Please Print in Full

Address: Malaga. Spain

City

Country _____ Nationality: Spanish. Car Registration Number _____
Passport No. (Aliens only) 20170E Where Issued _____
Next Destination (Give address) _____

ACCOUNT CUSTOMERS ONLY – If the Company for whatever reason fails to honour this account, I undertake to be held personally liable for the full payment thereof.

Receptionist: SW. Signature: P Martino

Card 4 (bottom right):

Length of stay	2N
Number in party	1

CASH ☐
CHEQUE ☐
A/C ☐
CREDIT CARD ☑
VOUCHER ☐

EROS HOTEL
65-73 Shaftesbury Avenue
London W1V 8EX
Telephone: 01-734 8781
Telex: 892676 VICGDN

MR. MRS. MISS: JONES (JSE) (Real Name Hamson Ford)
Please Print in Full

Address: BEVERLY HILLS
HOLLYWOOD. USA.

City

Country _____ Nationality _____ Car Registration Number _____
Passport No. (Aliens only) _____ Where Issued L.A
Next Destination (Give address) L.A

ACCOUNT CUSTOMERS ONLY – If the Company for whatever reason fails to honour this account, I undertake to be held personally liable for the full payment thereof.

Receptionist: SW. Signature: B Jones

Card 1 (top left)

Length of stay: 2 N	CASH ☑
	CHEQUE ☐
	A/C ☐
Number in party: 1	CREDIT CARD ☐
	VOUCHER ☐

EROS HOTEL
65-73 Shaftesbury Avenue
London W1V 8EX
Telephone: 01-734 8781
Telex: 892676 VICGDN

MR. / MRS. / MISS: Esra GUNENSTUTAR
Please Print in Full
Address: c/o HILTON HOTEL
ISTANBUL.
TURKEY. City
Country: Turkey. Nationality: TURKISH Car Registration Number: N/A
Passport No. (Aliens only): 79218 Where Issued: LONDON
Next Destination (Give address): Four Season Hotel. Toronto.
ACCOUNT CUSTOMERS ONLY – If the Company for whatever reason fails to honour this account, I undertake to be held personally liable for the full payment thereof.
Receptionist: SW Signature: E Smith

Card 2 (top right)

Length of stay: 7 N	CASH ☐
	CHEQUE ☐
	A/C ☐
Number in party: 2	CREDIT CARD ☐
	VOUCHER ☑

EROS HOTEL
65-73 Shaftesbury Avenue
London W1V 8EX
Telephone: 01-734 8781
Telex: 892676 VICGDN

MR. / MRS. / MISS: ANGELINO Sergio + Carla.
Please Print in Full
Address: 2. Calle Verona
ROMA- ITALIA.
City
Country: ITALY. Nationality: Italian Car Registration Number: —
Passport No. (Aliens only): 20577. 20611 Where Issued: ROME
Next Destination (Give address): ROME
ACCOUNT CUSTOMERS ONLY – If the Company for whatever reason fails to honour this account, I undertake to be held personally liable for the full payment thereof.
Receptionist: SW. Signature:

Card 3 (bottom left)

Length of stay: 3 N	CASH ☐
	CHEQUE ☑
	A/C ☐
Number in party: 2	CREDIT CARD ☐
	VOUCHER ☐

EROS HOTEL
65-73 Shaftesbury Avenue
London W1V 8EX
Telephone: 01-734 8781
Telex: 892676 VICGDN

MR. / MRS. / MISS: WALKER & JAMES.
Please Print in Full
Address: 76. Orchard Road,
Kibworth Beauchamp.
City
Country: Leics. Nationality: BRIT. Car Registration Number: C-652 TAY
Passport No. (Aliens only): Where Issued:
Next Destination (Give address): Home
ACCOUNT CUSTOMERS ONLY – If the Company for whatever reason fails to honour this account, I undertake to be held personally liable for the full payment thereof.
Receptionist: SW. Signature: J walker

Card 4 (bottom right)

Length of stay: 1 N	CASH ☐
	CHEQUE ☐
	A/C ☑
Number in party: 2	CREDIT CARD ☐
	VOUCHER ☐

EROS HOTEL
65-73 Shaftesbury Avenue
London W1V 8EX
Telephone: 01-734 8781
Telex: 892676 VICGDN

MR. / MRS. / MISS: BRITTAIN John + Susan,
Please Print in Full
Address: BOSTON. Lincs.
City
Country: Nationality: BRIT. Car Registration Number: A 421 BEA
Passport No. (Aliens only): Where Issued:
Next Destination (Give address):
ACCOUNT CUSTOMERS ONLY – If the Company for whatever reason fails to honour this account, I undertake to be held personally liable for the full payment thereof.
Receptionist: SW Signature: J B

Tasks

1 Check the cards.
2 List the names of the clients whose cards are ready for filing.
3 Explain in detail what is wrong, if anything, with the remaining cards.

30 minutes

3 Safe deposits

The Scenic Hotel is a 450 bedroom, city centre hotel attracting the business/tourist sector of the market.

The Riverside is a 28 room, 18th century establishment, attracting conference business and a good share of the domestic holiday market.

Both operate safe deposit systems for the security of valuables. The two receipts shown here are owned by two different guests, one of whom is staying at the Scenic and the other at the Riverside.

Tasks

1 State which receipt is likely to belong to which hotel and explain why.
2 Describe the procedure you would follow when both of the guests want their property back; Mr Kirby to take out some travellers' cheques, and Mrs Wells to put in some jewellery.
3 Outline how you would deal with an enquiry for the temporary safe-keeping of valuables from a gentleman who is a regular customer in your restaurant, but not resident in your hotel.
4 Outline how you would deal with Mme Guichteau who wishes to leave four items from her summer fashion collection in safekeeping.
5 Make the appropriate entries when Mr Kirby is clearing his safe deposit box and leaving the hotel.

45 minutes

Box No.				Room No.
Date	Time	I have this day opened my Safe Deposit Box and found everything in good order. *Signature of Guest :-*	In the presence of Cashier. *Signature :-*	

I have this day surrendered my safe deposit box in good order and have reclaimed all that was due to me

Date

Signature of Guest

H & B 7/76   Signature of Witness

No **7300**   DEPOSIT RECEIPT   No **7300**

.... 19 .... 19 ....

Article(s) ....   Name ....

....   Room No. ....

....   Article/s ....

....

....

Received the above article(s)

....   Receptionist's Signature:— ....

....

Guest's Signature   Guest's Signature:— ....

4 Bedsheets

Refer to the diary in Chapter 2, Assignment 11 (p. 30)

Tasks

1 Using the bedsheet, allocate rooms for the arrivals on 1st March – give reasons why you have selected particular rooms if appropriate.

2 Compile an arrivals list for 1st March.

3 List the departments to whom this list would be circulated.

4 By 22.00 hrs everyone has arrived except Holmes and the Katherens – compile a guest list.

5 List the departments who would need a copy of this list.

6 Compile a departure list for 2nd March, and state which departments would be in receipt of this list.

7 What information should be left for the Night Porter prior to leaving at 23.00 hrs.

During the night the Katherens arrived – the porter took a deposit for £100 and gave them room 11. Mr Holmes did not arrive.

Megatours telephoned to give final group numbers: 14T, 12S, 4D

The CRO have agreed rates of £30 – S, £40 – T, £40 – D, regardless of whether rooms are with or without bathrooms.

8 Bring the records up-to-date.

9 Explain how you would deal with the reservation for Mr Holmes.

10 Students who have access to a typewriter should type an Arrivals/Departure Guest List in accordance with normal hotel procedure.

1½ hours

Nightly Sheet

...................................Day   The Night of ..../......................./.......19.................

TYPE KEY: S = SINGLE   T = TWIN   TR = TREBLE   D = DOUBLE   B = BATH

Room No.	Type	Remarks	Guest's Name	No. in Room	Nightly Charges	No. of Nights	Room No.	Type	Remarks	Guest's Name	No. in Room	Nightly Charges	No. of Nights
1	S	B					4	TB	F C3				
2	S	B					7	TB	F C6				
3	S	F C4					16	TB	F C15				
6	S	F C7					19	TB	F C20				
8	S	F					30	TB	B				
10	S	B	WILLIS MR P.	1	30 00	2	33	TB	F				
12	S	B					34	TB	F				
14	S	B					36	TB	B				
15	S	F C16					37	TB	B				
20	S	F C19					40	TB	B				
22	S	B					43	TB	F C42				
			(A) BLOCK TOTAL				44	TB	F	CROWN M/M P.T.	2	55 00	2
							46	TB	B	BLYTHE M/M W.	2	55 00	2
17	SB	F					47	TB	B				
18	Ssh	F	JONES Ms B.	1	40 00	1	50	TB	B	JOHNSTON M/M B.	2	55 00	1
35	SB	B					51	TB	F				
45	SB	B					54	TB	F				
53	SB	F					56	TB	B				
55	SB	B					57	TB	B				
60	SB	B					66	TB	B				
61	SB	F					67	TB	B				
65	SB	B								(E) BLOCK TOTAL			
			(B) BLOCK TOTAL										
							11	D/SB	B				
5	D	F					31	TRB	F				
9	T	B					41	TRB	F				
21	D	B								(F) BLOCK TOTAL			
			(C) BLOCK TOTAL										

COMMENTS

A	
B	
C	
D	
E	
F	

Room No.	Type	Remarks	Guest's Name	No. in Room	Nightly Charges	No. of Nights
23	DB	B				
32	DB	F				
42	DB	F C43				
52	DB	F				
62	DB	F				
63	DB	F	COTTON M/M Y.	2	55 00	1
64	DB	F				
			(D) BLOCK TOTAL			

TRMS

No. of Rooms Occupied

No. of Guests

Control Signature: Checked by:

5 In-house problems

Refer to the bedsheet in Assignment 4 (p. 39).

Tasks

1 Chance arrival Mr and Mrs Green, two nights Room and Continental Breakfast. Allocate them a room and complete the key card below.

EROS HOTEL

65-73 Shaftesbury Avenue
London W1V 8EX
Telephone: 01-734 8781
Telex: 892676 VICGDN

Room Number	Name		
Room Type	Nightly Rate	No. of Guests	Departure Date

Please show this card in the restaurant in order to obtain your English breakfast.
Veuillez presenter cette carte au restaurant pour obtenir votre petit déjeuner anglais.

EROS HOTEL

69 SHAFTESBURY AVENUE LONDON W1V 8EX
Telephone: 01-734 8781 Telex: 268564 EROSHT

Room Number	Name		
Room Type	Nightly Rate	No. of Guests	Departure Date

Please show this card in the restaurant in order to obtain your Continental breakfast.
Veuillez presenter cette carte au restaurant pour obtenir votre petit déjeuner continentale.

```
REMOVAL SLIP -                          RECEPTION

Name ...........................................

From Room No ...................................

To Room No .....................................

Group Name .....................................

                        NO PERSONS

From ...............    To .....................

                        RATE CHANGE

From ...............    To .....................

Remarks:

Date ...............    Time ...................

Signature ......................................

                                       Reception
AH52
```

2 The occupants of room 44 report that the TV has no sound, although the picture is alright. The housekeeper confirms that this is so.
   (a) move the guest to another room;
   (b) send a memo to appropriate department regarding TV;
   (c) complete the move notification and list which departments should receive copies.

3 Mrs Filmer arrives. She has her copy of a confirmation letter for a single room for six nights from tonight. Her name does not appear on the arrival list. Explain in detail how you should deal with this situation.

4 The housekeeper urgently needs room 67 for window repair. Take off for two days.

5 Captain and Mrs I Arichi are arriving at 4.00 am tomorrow and paying for tonight. Pre-allocate a room.

1 hour

Double booking

Mr P Barker arrives with a confirmed reservation for a single for three nights.

Tasks

1 Itemise the check-in procedure and allocate him a room.

The Junior Receptionist escorts Mr Barker to his room (102) but on arrival it is found that the room is already occupied. The receptionist asks the resident guest his name and he replies 'Mr Barker'. She then returns to reception with Mr Barker number two.

2 Give a detailed explanation as to how you would investigate and rectify the situation, bearing in mind the hotel is full, except for a two bedroom suite at a considerably higher cost than the single Mr Barker had originally reserved.

30 minutes

6 Room status

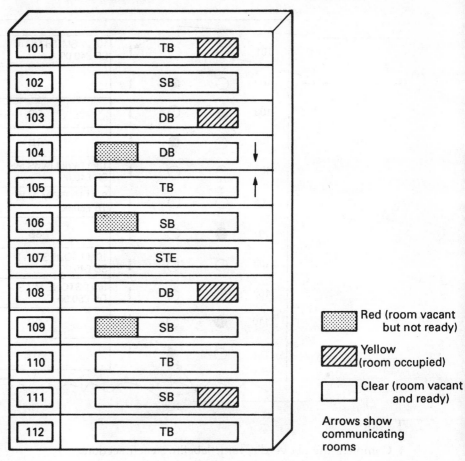

Room	Type	Status
101	TB	Yellow
102	SB	Clear
103	DB	Yellow
104	DB	Red (arrow down — communicating)
105	TB	(arrow up — communicating)
106	SB	Red
107	STE	Clear
108	DB	Yellow
109	SB	Red
110	TB	Clear
111	SB	Yellow
112	TB	Clear

Key:
- Red (room vacant but not ready)
- Yellow (room occupied)
- Clear (room vacant and ready)

Arrows show communicating rooms

Tasks

Look at the room status rack shown above.
1 If Mr and Mrs Brown arrive and request a double with private bathroom:
   (a) is there one available?
   (b) if so, when can they occupy the room?
2 How many rooms are occupied and what type of accommodation are they?
3 Ms J Smythe arrives and requests a single room with private bathroom:
   (a) is there one available?
   (b) if so, which room will you allocate?
4 The housekeeper telephones to advise you that room number 109 is ready.
5 List the status of each room after taking the above into consideration.

20 minutes

7 Electronic room board

Room				Details
101	○	○	▮	10/11 SMITH MR J 13/11 SB £40.00
102	○	●	▮	TB
103	●	○	▮	DB
104	○	●	▮	TB
105	○	○	▮	12/11 WILLIAMS MS 13/11 SB £40.00
106	○	○	▮	9/11 HYOE M/M W 14/11 TB £60.00
107	●	○	▮	SB
108	○	○	▮	10/11 JONES MR P 14/11 SB £40.00
109	○	○	▮	9/11 STONE MS S 13/11 DB £50.00
110	●	○	▮	DB
111	○	●	▮	STE

Tasks

1 Complete the above diagram, labelling each section.

2 List the status of each room.

3 How many rooms are occupied?

What type of accommodation are they?

4 Mr and Mrs P Stubbins arrive and request a double room with private bathroom:

  (a) is there a room available?

  (b) if so, which number, and when can they occupy the room?

5 Mr K Leedham arrives and requests a suite:

  (a) is there one available?

  (b) if so, which room will you allocate?

The housekeeper 'jacks' a room back for letting (number 103).

Ms Williams pays her account and leaves the hotel.

List the status of each room after taking the above factors into consideration.

30 minutes

Whitney Room Status Rack for 20/2

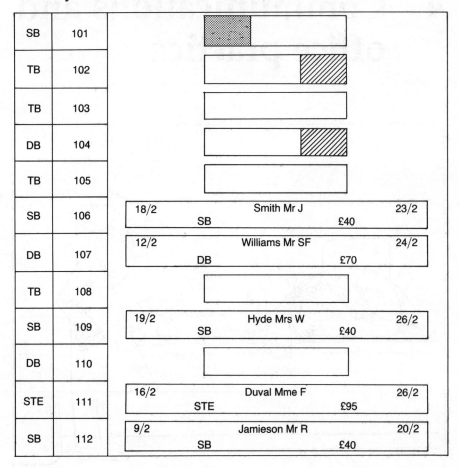

SB	101
TB	102
TB	103
DB	104
TB	105
SB	106
DB	107
TB	108
SB	109
DB	110
STE	111
SB	112

Room slips shown in rack:
- 106: 18/2 — Smith Mr J — 23/2 — SB — £40
- 107: 12/2 — Williams Mr SF — 24/2 — DB — £70
- 109: 19/2 — Hyde Mrs W — 26/2 — SB — £40
- 111: 16/2 — Duval Mme F — 26/2 — STE — £95
- 112: 9/2 — Jamieson Mr R — 20/2 — SB — £40

Tasks

1 List the status of all the rooms.

2 Mr and Mrs Stubbins arrive and request a twin room – is there one available? if so make the appropriate allocation.

3 Mr K Leedham arrives and requests a single room – you explain that you do not have one available. Sell him something as an alternative.

4 Mr and Mrs Wilson arrived early this morning and were allocated room 104. They are staying for four nights, and sending their account to Wilson Wanderers (booking was made by telex). Draw out the shannon as it would appear in the rack.

5 Mr Jamieson pays his account and leaves the hotel.

6 List the status of each room after taking these factors into consideration.

20 minutes

4 Communications and office practice

1   Verbal communication

The following situations arise while you on duty at reception:
- the housekeeper wants the maintenance man
- the butcher is delivering turkeys for the chef, at the goods inwards entrance
- there is a telephone call for Mr Valera, Room 305
- the husband of Mrs Worthington is at reception
- telephone message for Lyn, courier of Excell Tours
- security officer wanted, to attend to trouble in the bar
- John, the porter's girlfriend, is on the phone

Task

State what you would say in each of those situations when using the public address system.

45 minutes

The banqueting office has an external telephone line, which is open between 9.00 am and 6.00 pm. An answer phone has just been purchased, for use when the office is closed.

Tasks

1 Write a suitable message to be used for the recorded message on the answer phone.
2 Record your message and discuss any weak points with the others in your group, and how it might be improved.

1 hour

2 Postal services

The following items must be despatched today. The hotel makes use of all of the services listed.

- Recorded delivery
- Registered post
- Proof of posting
- Business Reply Service
- Red Star
- Air Mail
- surface mail
- fax
- telex
- Telemessage

(a) guest in Room 204 has an important message for her company in Australia

(b) guests' lost property (a watch) to be returned

(c) brochure and conference details to US travel agent

(d) sales figures to the General Manager, who is at Head Office in a meeting.

(e) copy of disciplinary letter sent to Banqueting Manager, to Head Office Personnel

(f) box of company computer stationery to the sister hotel in Leeds

(g) information on deep fat fryer for Chef, as seen in the *Caterer and Hotelkeeper*

(h) guest wants letter sending to his solicitor confirming acceptance of offer

(i) congratulations to Front Office Manager on the birth of her baby

(j) brochure and tariff to prospective client in Bath

Tasks

1 Which would be the most appropriate to use in each case?
2 State approximately how long each would take to reach their destination.

45 minutes

3 Letter writing

Refer back to Assignment 6 on accommodation availability, Chapter 2, page 20. There were a number of clients on that occasion that you were unable to accommodate.

Task

1 Write to *one* of those clients using the letterhead below, informing them that you are unable to help them.
2 Write a second letter offering them some kind of alternative.

The letters should be properly laid out in a business-like fashion – use your own imagination for details of names, addresses etc. If you have access to a typewriter, type the letters in keeping with normal hotel procedure.

45 minutes

Basil Street Knightsbridge London SW3 1AT
Telephone 01-589 5171 Telex 919042
Cables/telegrams Hotelcap LDN

CAPITAL HOTEL

4 Telephone meters

The following rooms are due to depart and you have been asked to prepare the telephone charges from the meters located in the cashier's office. The readings are as follows:

Room	Reading
101	02435
201	01459
315	11782
410	12212
115	05782

The previous readings had been noted on the respective bills and were as follows:

Room	Reading
101	02418
202	01421
315	11680
410	12190
115	05770

The hotel charges 20p for the first unit and 12p for subsequent units used. The current British Telecom rate is 4.4p per unit plus VAT.

Tasks

1 Show the charge incurred by each room for telephone and VAT.
2 Using the data provided state how much profit the hotel has made.

45 minutes

5 Telecommunications

Room	Time	Name	Units	Price
109			18	
321			7	
111			12	
110			32	
181			2	
301	0905	Lawrence	N/R	
321			2	
321			3	
209			8	
212	0950	Firth		£4.12
119			17	
111			3	
112			9	
311	1115	Walsh		£0.95
317			41	
221			132	
114	1425	Freeman		£1.50 (reverse charge)
210		Willis	4	
304	1505			£3.70

House calls

	Time	Units
Advance reservations to Paris	1305	25
Banqueting to Scotland	1330	18
Accounts to Sunderland	1345	48

- ADC calls are subject to 50% surcharge
- house calls are priced at the British Telecom rate of 4.4p per unit
- guest units are charged at the rate of 15p per unit

Tasks

1 Enter the telephone calls from your shift onto the summary sheet provided.
2 During your shift how many units have been used?
3 How much is the hotel being charged by British Telecom
4 How much revenue is earned from the guest units?
5 Has the hotel made a profit this shift?

45 minutes

TELEPHONE SUMMARY SHEET

Date .... Shift ....

Operator ....

Room No.	Time	Name	Comments	Units	Guest Calls £	Guest Calls p	House Calls £	House Calls p	Operator Connected Calls £	Operator Connected Calls p
TOTALS										

6 Star ratings

Divide into groups of 3–4 students.

Tasks

1 Choose a number of hotels in your locality – enough to make a comparison. If you live near a large city you may be able to compare new hotels with old hotels, or large hotels with small hotels.

Look up the telephone numbers of the hotels you have selected. Each member of your group should ring at least one hotel. The suggested conversation (although you may have your own ideas) is:
- as soon as the telephone is answered – before it can be transferred – say 'have you got a room for the night of . . .?'
- as soon as the call is transferred say 'I am getting married soon and would like to book a room' – let the staff suggest a room type and try to sell it to you
- ask about dinner – what time, price etc.
- ask about parking and directions to the hotel
- ask for a brochure and tariff to be sent to you.

While you are ringing make a note of the following points:
(a) the length of time the switchboard take to answer
(b) how the telephone is answered – was the name of the hotel given? etc.
(c) how long it takes to transfer the call to the relevant department
(d) whether the call is transferred without any problems
(e) how long the department takes to answer
(f) the attitude of the reservations staff, their product knowledge etc.

After the call, remember to note how long it takes for the tariff/brochure to arrive, and the standard of the stationery accompanying it.

2 A few days later try a business-like approach, knowing your dates prices etc. and see if the attitude of the staff changes.

3 Draw up a chart of all the hotels you contacted and compare them. Give them your own star ratings.

7 Filing

Your company is organising a sales drive aimed at repeat business throughout the UK.

The computer has produced a list of everyone who has stayed only once at the hotel during the last 18 months and the name of the town in which they live The country has been divided into six regions for the purpose of the sales drive.

Tasks

1 Arrange the towns according to their geographical region. Use an atlas to help you.

2 File the clients' names in alphabetical order within their geographical regions.

1½ hours

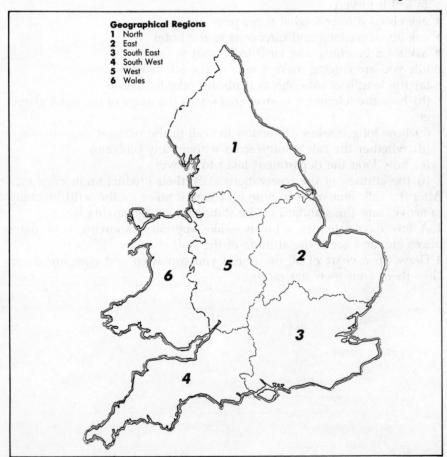

Geographical Regions
1 North
2 East
3 South East
4 South West
5 West
6 Wales

Macclesfield	G Vanetzian	Birmingham	J A Smith
Lincoln	I Niknam	Worcester	C Faulkner
Dover	G Barker	Barnsley	M Bate
Milton Keynes	A Van Reysen	Swansea	G Smith
Glasgow	G McKie	Grimsby	C Noakes
Oxford	K Maynard	Reading	D Knight
Blackpool	S Wilson	Bournemouth	G Arkless
Cardiff	L Davies	Sunderland	P Glasgow
Stoke-on-Trent	H Davis	Crewe	F Mellon
Taunton	T Barker	Cardiff	P Smithe
Aberdeen	M MacFarlane	Aylesbury	S Seward
Kendal	J Smith	Bognor	C Granger
Bradford	G Ray	Plymouth	A Bates
Great Yarmouth	A Barrett	Bristol	R Noakes
Carlisle	B Reynolds	Norwich	C Killingworth-Baird
Manchester	A Astles	Coventry	R Joseph
Exeter	G Careless	Carlisle	I Gupta
Basingstoke	S DeLissandri	Scunthorpe	L Carter
Malvern	C Ritchie	Aberystwyth	J Sheppard
Sheffield	J Cooper	Colchester	H Clark
Maidstone	E Rosner	Birmingham	C Wade
Ayr	T Gough	Brighton	S Wright
Peterborough	M Legg	Chester	D Murphy
Hastings	R Williams	Mansfield	G Clarke
York	T Nutt	Cambridge	T Waide
Hereford	M Connolly	Newcastle	A Bradley

8 Job applications

Tasks

1 Complete the application form to apply for the vacancy advertised.
2 As a group discuss under the following headings, the person most likely to get the job.
- qualifications
- experience
- references
- social skills
- appearance
- acceptability

1½ hours

SURNAME		CHRISTIAN NAMES		MR MRS MISS
ADDRESS			HOME TELEPHONE	
DATE OF BIRTH	NATIONALITY		MARRIED/SINGLE	NO. OF CHILDREN
NEXT OF KIN	POSITION APPLIED FOR		CURRENT SALARY	DATE AVAILABLE FOR EMPLOYMENT IF OFFERED

HEALTH – (All senior employees are required to pass a Medical examination Please state if you have any physical handicaps or have had any serious illnesses.)

EDUCATION – Schools attended	PERIOD

Scholarships gained, Degrees, Diplomas or Professional Qualifications. (Please give dates.)

Please show here any additional information about yourself you would like us to know, e.g. hobbies, interests, ambitions, membership of Clubs, societies and other bodies.

LANGUAGES	SPEAK			WRITE		
	Fluent	Good	Fair	Fluent	Good	Fair
ENGLISH						
GERMAN						
FRENCH						
SPANISH						

EMPLOYMENT RECORD

Name and address of employer (Last employer first)	From	To	Position and salary With/without accommodation	Reason for leaving

WELLAND HOUSE HOTEL

JUNIOR RECEPTIONIST

Required for busy 72 bedroom City Centre Hotel
The successful applicant will be aged
18 or over, with smart appearance and
a friendly manner. Responsibilities
will include general reception and cashier
duties, reservations and typing. Five
day week, straight shifts, some weekends.
The position offers a good salary, live in if
required.

Please telephone Mrs Heeble, 0205 60123 for
application form

57

9 Circulating information

Mrs Wilmer Katz arrives at the hotel at 10.30 to take up the reservation made on behalf of herself and her husband. Mr Katz has gone straight to the offices of his UK company for an important meeting.

The porter shows Mrs Katz to room 112. About 20 minutes later she phones down to reception to complain – the room is noisy and very pokey she says, and she would like something better. You are very busy so you ask the porter to show her 214 and 218 as alternatives. When he comes down he says she has half unpacked and used the bathroom and he wouldn't move her if it were left to him.

Mrs Katz phones down about 10 minutes later and says she would like to take 218, but she would like some help moving her belongings. You ask the housekeeper who is not very pleased since she is short of staff. A house porter is eventually sent.

Twenty minutes later Mrs Katz appears in a hysterical state saying her gold and diamond watch has gone and that it must have been stolen. You ring the housekeeper and ask her to check both rooms while you try and calm Mrs Katz. She phones her husband but his office receptionist tells her he is very busy and cannot be disturbed, but he will call back as soon as possible. Mrs Katz goes to lie down and takes a tablet and asks you to be sure to take a message from her husband.

She returns to the desk an hour later – there is no call from her husband and she is starting to become hysterical again. You suggest she goes for a short walk to get some fresh air, and you offer to call her husband again. She gives you the number and goes out. When you get through to his company you get connected to Mr Katz who is absolutely furious. He attempted to call his wife once already and was put through to room 112 to a stranger. You realise you forgot to tell the switchboard that Mrs Katz had moved. Mr Katz is furious at the interruptions and becomes even angrier when you tell him what the problem is. He leaves you in no doubt at all what he thinks about the hotel and the situation, and furthermore begins an attack on his wife, who he says is always losing things and has got no sense of perspective.

When you leave at 3.00 Mrs Katz still has not returned.

Tasks

1 Complete the message form notifying Mrs Katz that you have made contact with her husband.
2 Send a memo to the Duty Manager to acquaint him with the situation.
3 Write a report for the next shift to put in the hand-over book.

1 hour

Telephone

Date _____   Time _____

Name _____

Address _____

_____

Phone No _____

Taken by _____

10 People and communication

You have a new job as a shift supervisor in a hotel/hostel which is very busy. The shifts are arranged in an unusual manner; mornings only, evenings only or nights. You are supervising the early shift. Shifts are from 7.00–3.00, 3.00–11.00 and 11.00 to 7.00 with a half-hour overlap by one shift member on a rota basis.

There are five members of your team, excluding yourself. Three of the staff have been with the company for some time (one had applied for your job as supervisor) one started two weeks ago, and one, a management trainee, is with the hotel for a one-year industrial release programme from university. She has spent two months in housekeeping and has been in your department for ten days. The atmosphere is rather tense.

The three well-established members have formed a 'clique', and seem to go out of their way to exclude the other two. They do little to help the two new staff members, who are still settling in. Jane, who began two weeks ago, has had experience in a similar operation, but is still learning the ropes, and Marion, the degree student has really only theoretical knowledge.

You are told by the late shift supervisor that you should do something about the bad atmosphere, since one or two of the guests have commented upon it. You know the management are keen on good staff relations, and you are worried in case it filters through to them.

You decide to call a meeting of the staff concerned at 3.30 pm and send a note to everyone to this effect, giving one week's notice. One of the established team says quite clearly, in your hearing, that you can count her out, stating she spends her free time as she pleases. The other two are off-hand, and say nothing one way or the other.

You are off duty for two days, and when you return you find the situation has worsened. Jane is very despondent all day, and after the main morning rush is over she says she would like to transfer to the late shift, or even to nights. When pressed as to why she starts to cry and says no-one ever tells her anything, the others expect her to know everything, and Marion is very rude and condescending. You send her home a bit early, and say you will speak to her tomorrow. At the end of the shift you ask to speak to Marion, and briefly recount what Jane has said. She seems to find it all, very amusing, and implies as she leaves that Jane is not very bright, and neither are you if you let something like that worry you.

The girl from your shift who is on hand-over duty is talking to a member of the late shift, and to your dismay you hear her say that their new shift supervisor is still 'wet behind the ears'.

Tasks

1 Outline the main reason for the problems.
2 Explain how some of them could have been avoided.
3 List ways in which a similar situation could be avoided in the future.
4 Describe what action you are going to take.

1 hour (including discussion)

5 Billing

R.B. JACKSON

1   Account preparation

Mr and Mrs Niknam have checked into room 401 at 3.00 am on the 26th December, and are staying until 29th December in a DB at Room Only Rate of £45.00 per night inclusive of VAT. They have paid a £50 cash deposit on arrival. The following charges are incurred by them.

Dec 26th
Breakfast	£7.50
Newspaper	.32p
Telephone	£2.92
Dinner	£17.00
Wine	£6.32
Telephone 10 units @ 12p per unit	

Dec 27th
Breakfast	£3.50
EMT	.90p
Newspaper	.32
Theatre Tickets	£16.20
Supper	£8.00
Drinks	£3.40

Dec 28th
Breakfast	£7.50
Packed Lunch	£7.00
Telephone	£2.50
Coffee	£1.20
Brandy	£2.80
21 Telephone units	

Dec 29th
EMT	.90
Breakfast	£7.50

Tasks

1 Enter the charges on Mr and Mrs Niknam's bill.
2 Enter VAT as appropriate.
3 Total the bill.

45 minutes

M_____ Room No._____

_____

DATE								
Brought forward								
Apartment								
Breakfast								
Early Morning Tea/Coffee								
Newspapers								
Telephone								
Paid Out								
Laundry								
Sundries								
Lunch								
Dinner								
Wines & Spirits								
DAILY TOTAL								
Service Charge								
GRAND TOTAL								
Less: Cash Received								
Deposit								
Carried Forward								

VISITORS ARE REQUESTED TO VACATE THEIR APARTMENTS BY 12 NOON
ON THE DAY OF DEPARTURE.
ACCOUNTS DUE ON PRESENTATION.   **PLEASE LEAVE YOUR KEY.**

2 Tabular ledger – 1

The transfer of charges to a guest's account plays an important part in the business of a hotel. Most systems enable the management to see at a glance what the income/revenue is for each department, and what the daily takings are. The most well-established method of recording charges is the tabular ledger, still widely used in many hotels.

Even for those hotels whose systems are mechanised or computerised the principle remains the same. It is therefore essential for a receptionist to be able to complete a tab, of any sort, easily and efficiently.

Tasks

The following six assignments are arranged in order of difficulty.
There are two types of tabs – vertical and horizontal. Try the assignments on *both* layouts.

1–2 hours

TABULAR DAILY REPORT & CONTROL OF BUSINESS DONE

DAY OF WEEK ....

DATE ....

Inv. Number	GUEST'S NAME	Room No.	Sleepers	DAILY CHARGES								Daily Total	Balance B/F from Previous day	Grand Total	Cash Received	Ledger Received	Accounts Transferred to Ledger	Carried Forward
				Room	Breakfast	Food	Dinner	Bar	Wines	Telephone	Other							

TOTAL

SLEEPERS

CANDIDATES NO.

DATE

DAILY TOTAL

ROOM NO.
NAME
RATE
B/F
APARTMENTS
PENSION
BREAKFASTS
LUNCHEONS
TEAS
DINNERS
EARLY TEAS
BEVERAGES
WINES
SPIRITS & LIQUEURS
BEERS
MINERALS
TELEPHONES
V.P.O'S
NEWSPAPERS
TOTAL
CASH
ALLOWANCES
LEDGER
BALANCE C/F

PM/J/1090

1 The Sunny Hotel in Rainzalotte offers its clientele two types of tariff, as you can see:

Room and Breakfast (R & B) at £38.00 per person
Room Only (RO) at £30.00
EMT – 40p, Coffee 50p
English breakfast (E) £4.00
Continental breakfast (C) £1.50
Lunch £5.50
Dinner £8.50
Afternoon tea £1.60 per person
Room and Breakfast rate includes full English Breakfast.

Room	Name	Terms	B/F
101	Mr/s Wilson	RO	£90.59
103	Mrs Wales	R & B	£13.80
104	Mr Sinclair	RO	£68.09
106	Mr/s Scoot	R & B	£90.60

07.30 EMT to all residents.
08.00 Newspapers: **101** 80p, **103** £1.20, **106** 45p.
08.30 Breakfasts: 101 1E, 1C, **103** 1E, **104** 1C, **106** 2E.
09.00 Mr Welland arrives and is allocated Room **102** on R & B terms. He has English breakfast.
10.00 Telephone: **101** £5.90, **103** £1.20, **102** 56p.
10.30 Mrs Wales checks out. She pays her account by cheque.
11.00 Dry cleaning returned to **101** £6.80, **106** £4.70.
11.30 Mr Wilson gives you £100 cash on account.
12.00 Mr/s James and their baby arrive – Room **107** R & B terms, with a cot; £2.00 extra for the cot.
12.30 Room service: sandwiches and wine to Room **107** £3.20 and £5.60.
13.00 Lunch: **101** 2 covers, wines £5.70, 2 coffees, **102** 2 lagers £2.40, coffees, **107** 1 cover, wine £3.50, brandy £2.90.
15.00 Guest in room **104** wants to change room; he is moved to **110**, same rate.
15.30 Afternoon teas: rooms **101** 2, **106** + 2 guests, **110** 1.
18.00 Bar: **101** 2 gin and tonics, £4.80, cigarettes £1.90.
20.30 Dinner: **101** 3 wines £8.40, 2 cointreaux £4.80, coffees, **107** 2 wines £3.20, coffees.
21.30 **107** have hot milk (75p) for the baby.
22.00 Telephone: **101** £5.80, **107** £2.60.
23.00 Balance the totals and carry forward to the next day.

2 The Sunny Hotel – Day 2. Bring forward all the residents from the previous day, re-arranging their room numbers into the correct order:

07.30 EMT to all residents.
08.00 Breakfasts: **101** 2E, **107** 1C, 1E, **110** 1E.
08.30 Newspapers: **102** 40p, **107** 60p, **110** £1.00.
09.00 Telephone: **106** 90p, **107** 28p.
09.30 New arrival, Mrs Elliott. R & B terms **104**. She has English breakfast and a guest has Continental.
10.00 Room **102** is leaving, pays with a traveller's cheque.
10.30 Mr James pays £75 cash towards his account.
11.00 Flowers are delivered to room **101** £4.80, room **110** buys theatre tickets £4.90.
11.30 Bar: **101** 2 whiskies £2.80, cigarettes £1.90, **110** 2 sherries £1.18.
12.00 Lunch: **101** 2 covers, wine £3.40, coffees, **106** 1 cover, Pale Ale £1.20, **107** 3 covers, wine £6.80, liqueurs £2.20, 3 coffees.
14.00 Taxi fare **106** £1.40.
15.00 Afternoon tea for all.
15.30 New arrival, Mrs Wilkins. **102** RO.
19.30 Dinner: **101** 1 cover, Guinness £1.80, **102** 3 covers, champagne £18.00, coffees **107** 2 covers, wine £4.20, coffees.
22.00 Night porter sales: **110** sandwiches £4.80 and hot chocolate £1.00, **107** hot milk £1.00.

Complete the tab for the day, and balance.

3 Parisienne Hotel terms.

Room and Breakfast (R & B) £45.00 per person (inclusive of £5.00 breakfast)
Room only (RO) £40 per person
Inclusive £70 per person (inc. of £40 for board, which includes afternoon tea)
Morning tea – 75p (inc. of biscuits)
Beverages – 50p with meals
Lunch £9.50
Dinner £15.50

Room	Name	Terms	B/F
1111	Mrs T Thompson	Inc	160.56
1112	Miss Wilkinson	R & B	180.29
1113	Mr/s N Newman	Inc	101.57
1114	Mr/s Jenkins	Inc	210.76
1115	Miss P Perkins	R & B	20.00CR
1116	Mr R Roberts	R/O	152.90

07.00   EMT to all residents.

08.00   Breakfast: all residents, except **1114**; **1116** has one guest.

09.45   Newspapers: rooms **1111, 1113, 1114**   40p each.

10.00   Laundry returned: **1115**   £6.00, **1116**   £4.00.

10.30   Telephone calls: **1112**   £2.27, **1114**   £5.01.

11.30   Mrs Jenkins leaves and pays full bill to date in cash.

12.00   Lunches: **1111**   1 cover, wine £7.20, 2 coffees
        1113   4 covers, wine £12.50, beers, £3.00,
        1116   2 covers, spirits, £6.25, 2 coffees.
        Chance sales: lunches £142.50, wine £65.00.

13.30   Telephone: **1114** (Mr Jenkins) £2.00.

14.00   Mr/s D Davies arrive: **1117** RO, 2 coffees, £1.00 and sandwiches 2, at £2.00 each.

15.00   Mr Jenkins and Mr Roberts leave. Mr Roberts' bill is paid by Jenkins Ltd.

15.30   Afternoon tea: all residents, £1.50 each.

17.00   Disbursements: **1113**, flowers £10, **1117**, postcards 70p, **1111** stamps 50p.

19.30   Dinners: **1113**   2 covers, minerals £2.50, 2 coffees, **1115**   3 covers, champagne £46.00, 3 coffees, **1117**   2 covers, wine £8.50, 2 coffees, liqueurs £4.00.
        Chance dinners: £248, wines £123.00, beverages £57.00, spirits £48.90, minerals £9.20.

20.00   Chance telephone £7.50.

23.00   Enter the next day's terms, and complete the tab for the day.

4 The King William Hotel Terms:

R & B £34.50 pp (inc. of £4.50 Breakfast)
RO £32.00 per person
Inclusive £57.00 per person (inc. of £34.00 board)
Private bathroom £4.00 per room per night
English breakfast £4.50
Continental Breakfast £2.25
EMT 70p per person
Morning coffee 75p per person
Lunch £9.00 per person
Meal beverage 75p per person
NB Terms are charged in this hotel in advance, or for new arrivals immediately upon arrival.

Room	Room	Name	terms	B/F
S	141	Miss M Timms	Inc	117.90
DB	151	Mr/s P Hunt	R & B	85.75
T	154	Lord & Lady Wellar	R & B	227.18
SB	161	Mr W Searle	RO	101.98

07.30	EMT to all residents.

07.30   EMT to all residents.
          Newspapers: **141** 57p, **151** 48p, **161** 28p, **154** 90p.

08.00   Telephone: **161** 93p, **154** £2.90.

08.15   **141** Flowers £5.00.

08.30   Chance breakfast £9.45, **151**-1E 1C, **161**-2E, **154**-1E 1C, **141**-2E.

09.00   **161** leaves and pays the account by cheque.

09.30   **151** has a cash advance of £50.

10.00   Miss Timms pays £90 on account with travellers' cheques.

11.00   Morning coffee to all residents.

12.15   Mr/s Miller arrive and are given room **181** on inclusive terms – DB.

12.30   **141** 1 lunch, 1 coffee, 1 brandy £1.45, **151** 1 lunch £8.50, coffee 80p, 1 bottle of claret £8.30, cigarettes £1.36, **181** 3 lunches, 2 coffees £1.50, lager £2.48, cordials 49p. Chance lunches £126.00, beverages £22.40, wines £70.90, spirits £106.70, minerals £4.37, cigarettes £12.50.

13.30   New arrival: Mr Jay, RO, **166** – SB.

14.00   Private lunch for the National Piggy Bank
          26 lunches, £12.50 each, inclusive of £2.00 pp for room hire
          26 coffees, 40p each, 26 brandies, 90p each, wine £124.30. The account should be sent to ledger – A/C No. S/19b.

14.30   Change of room **141** goes to **171** – SB.

14.45   Control inform you of the following errors:
          Room **151** was charged in error for 1 EMT instead of **141**; Room **151** was undercharged for dinner by 85p.

15.00   Cocktail bar pays in £87.54 liquor, £8.60 cigarettes.

17.00   £98.65 paid in by the telephone supervisor after emptying the kiosk. Post the apartments and close the tab.

5 The Panarama Hotel

R & B	£14.00 (inclusive of £2.00 breakfast) per person
Inclusive	£24.00 (inclusive of board £12.00) per person
Private Bathroom	£8.00 per room per night
EMT	40p per cup
Morning coffee	60p per cup
Table D'hote Lunch	£4.00 per person
Meal beverage	40p per person

Type	Terms	Room	Name	B/F
DB	Inc	201	Mr/s Gutteridge	49.80
TB	R & B	202	Mr/s Murphy	99.20
SB	R & B	203	Miss Herbert	81.30
S	Inc	204	Mr Kumar	18.45
SB	R & B	206	Miss Reeves	80.19
SB	R & B	207	Lady Wood	101.70

07.30	EMT to all
	Newspapers: **201** 28p, **202** 30p, **204** 28p, **207** 23p.
07.45	Telephone calls: **202** £1.90, **203** £4.60, **206** £2.20, **207** £1.90.
08.00	**206** checks out and pays cash.
08.30	Theatre tickets for Lady Wood £18.90.
09.00	Chance breakfasts: £23.80.
09.30	Room **203** checks out – sends the account to French Fashions Ltd.
10.00	Mr Kumar moves to **206**.
10.30	Mr Murphy pays £50 on account.
11.00	Chance telephone: £9.80.
11.00	Morning coffee to all, plus chance £7.20.
12.15	Arrival Ms Read: **209** – S Inclusive.
12.30	Luncheon: **201** 1 Table d'hote, 1 coffee, wine £2.80, **202** 2 Table d'hote, 2 coffee, wine £4.80, **206** 2 Table d'hote, 2 coffee, wine £3.20.
	Chance: 24 Table d'hote, à la carte lunches £92.90, beers £18.40, wines £120.70, spirits £54.80, minerals £1.90, tobacco £4.86.
14.30	Allowances and adjustments: Room **202** was charged for wine yesterday in error – should have been **201**.
	Room **207** was undercharged telephone: 50p instead of £5.00.
15.00	Arrival: Mr/s Tudisca arrive – room **210**, TB on R & B terms. Mr Tudisca operates a travel agency business in Italy and the Manager is giving him a 10% discount on apartments – to be deducted each evening.
15.30	Farewell luncheon of the Ladies' Sea Angling Club – held in the Mallinson Room: 15 covers at £9.60 inclusive of 60p room hire; sherry for all the group at 45p per glass, except Miss Wilson and Ms Wells who receive a complimentary glass; wines £96.00, liqueurs £62.00, cigarettes £4.20.
	Account to be sent to Mr A Hussain for payment – Ledger Folio No, L/21.
16.00	Lounge bar pays in £231.30 for spirits and liqueurs, and £21.90 beers.

Balance and reconcile the totals as if it were the end of the day's business.

6 City Centre Hotel tariff

R & B £18.00 (inclusive of £2.50 for breakfast) per person
Inclusive £28.00 (inclusive of £18.00 board) per person
Room only £15.00 per person
Private bathroom £4.00 per room per night
EMT 40p per person, morning coffee 45p per person
Table d'Hote Lunch £4.00 per person
Dinner £8.50 per person
Meal beverage 50p per person

NB Terms are charged in this hotel in advance, or for new arrivals immediately upon arrival. Inclusive terms cover the basic elements for this type of tariff.

Type	Terms	Room	Name	B/F
S	Inc	141	Miss G Granshaw	17.90
DB	R & B	151	Mr/s J Tasker	36.90
T	R & B	155	Mr Wells and Mr Wilson	91.19
SB	R/O	158	Mlle Belmonte	18.97

07.30   EMT to all residents.
         Newspapers: **141** 27p, **151** 30p, **158** 20p.
08.00   Breakfast to all residents.
08.15   Flowers, **141**, £6.80.
08.30   Chance breakfasts 21.
09.00   Room **155** checks out and pays £50.00 on Access and transfers the rest to his company – a/c folio SL370
09.30   Room **151** has a cash advance of £30.00 – VPO.
10.00   Morning coffee to all residents.
         Chance morning coffees £8.20.
12.15   Mr/s J Whales arrive – they are allocated room **155** on inclusive terms.
         Mr R Bennett arrives – he is allocated **181** on inclusive terms – a single with bath. He is a chance guest and pays an advance deposit of £20 with travellers' cheques.
12.30   Room **141** 1 lunch, 1 coffee, brandy 90p.
         Room **151** 2 lunches, 2 coffees, 1 bottle of claret £7.20, cigars £2.90.
         Room **158** 2 lunches, 2 coffees, 2 beers 80p each.
         Room **181** 1 lunch, 1 coffee, cigarettes £1.10.
         Chance lunch: 18 Table d'hote, á la carte £56.80, beverages £5.70, wines £80.91, spirits £30.16, cordials £3.90, cigarettes £2.50.
14.30   Change of room, **141** moves to **171** – SB.

14.45 *Arrivals*

The sales and Marketing Institute book in six delegates to a conference being held in the hotel. They are sharing TB rooms **190, 192, 194**. The company will pay for their rooms (room only basis) and main meals. Open a main bill and extras for the rest of their charges:

190 – Misses James and Nichols

192 – Mr Willis and Mrs Seret

194 – Misses Johnson and Wilkins.

15.00 Control call: the following errors occured yesterday:

Room **151** was charged in error for 1 EMT – should have been **141**, Room **151** was undercharged for dinner – £1.30.

16.00 Rooms **190** and **194** take tea and sandwiches in the lounge – the total bill is £4.90.

17.00 Telephone calls: **181** £3.90, **190** £1.90, **192** 80p, **194** 30p.

18.00 Cocktail bar, Room **194** £4.90

21.00 Dinner is taken by the following:

158 3 dinners, 3 coffees, wine £6.90.

190/192/194 6 dinners, wine £18.90, cigarettes £3.30, liqueurs £6.60.

22.00 Cocktail bar pays in £80.90 spirits, £3.80 beers and £3.90 minerals. Balance the ledger for the day.

3   Taking over a tab

The tab ledger which you take over from the first shift is complete except for the entries listed overleaf and tonight's appartments.
(TB £65, SB £35)

Tasks

1 Mr Dilaver is obliged to leave suddenly owing to illness in the family – the hotel will not charge for tonight.
Make up his bill.
2 Explain how you will refund the money which is outstanding.
3 Complete and balance the tab overleaf, locating and correcting any errors if necessary.

30 minutes

M_____			Room No._____						
_____									
DATE									
Brought forward									
Apartment									
Breakfast									
Early Morning Tea/Coffee									
Newspapers									
Telephone									
Paid Out									
Laundry									
Sundries									
Lunch									
Dinner									
Wines & Spirits									
DAILY TOTAL									
Service Charge									
GRAND TOTAL									
Less: Cash Received									
Deposit									
Carried Forward									

VISITORS ARE REQUESTED TO VACATE THEIR APARTMENTS BY 12 NOON
ON THE DAY OF DEPARTURE.
ACCOUNTS DUE ON PRESENTATION.   **PLEASE LEAVE YOUR KEY.**

TABULAR DAILY REPORT & CONTROL OF BUSINESS DONE

DAY OF WEEK .... TUESDAY
DATE 4.8.6X

Inv. Number	GUEST'S NAME	Room No	Sleepers	Room	Breakfast	Food	Dinner	Bar	Wines	Telephone	Other @ (1 day)	Daily Total	Balance B/F from Previous day	Grand Total	Cash Received	Ledger Received	Accounts Transferred to Ledger	Carried Forward
0914	Walton Mr/s	111	2		7 00	12 40				8 20			112 51					—
0919	Tsar Mr	210	1		3 50		10 50		8 50				90 28	112 68			112 68	
0944	Dilaver Mr	212	1		3 50			5 50		4 60			151 41		200 00			
0946	Wilkins Mrs	217	1		3 50		10 50		7 75		@ 4.40		94 81					
0951	Wadsworth Mr	104	1												50 00			
0952	James Mr/s	119	2				17 00		12 95									
TOTAL																		

4 Tabular ledger 2

Using the following information, calculate today's balance for the tabular ledger sheet below. You are not required to make entries on the ledger sheet.

101 Pays cash on departure
102 Still in residence but guest exchanges a £20.00 travellers cheque for cash.
103 Still resident, but was overcharged last night 50p for a coffee.
104 Checks out and pays by American Express.
105 Departs with an advance deposit of £20.00 and pays the difference by Access.
106 Pays £50.00 cash on account.

Chance restaurant: £10.00 Visa, £15.00 Diners Club, remainder in cash.
IBP Conference: Account to   IBP Ltd
                               Ferry Lane
                               Wilborough

Chance breakfasts: £2.80 (cheque), remainder in cash.

Chance morning coffee: cash.

1 Hour

SLEEPERS _____   CANDIDATES NO. _____

ROOM NO.	101	102	103	104	105	106		Chance Restaurant	IBP Conference	Chance Brkfts	Chance AM coffee			DAILY TOTAL
NAME														
RATE														
B/F														76 35
APARTMENTS														53 00
PENSION														18 00
BREAKFASTS														4 30
LUNCHEONS														12 00
TEAS														1 40
DINNERS														25 50
EARLY TEAS														1 20
BEVERAGES														2 52
WINES														11 20
SPIRITS & LIQUEURS														—
BEERS														15
MINERALS														24 80
TELEPHONES														61
V.P.O'S														2 90
NEWSPAPERS														27
														21 40
TOTAL	20 88	23 80	32 22	22 35	30 70	9 85		39 00	34 90	18 10	3 80			
CASH														
ALLOWANCES														
LEDGER														
BALANCE C/F														

PS/L/3036

5   Machine accounts

Hotels have a wide variety of mechanised, electro-mechanical or computerised accounting systems. You may not be proficient in all accounting
methods but practice on one type of machine will be invaluable for the
future.

Following is a series of machine accounts exercises. Enter these charges
on a guest bill via an accounting system (e.g. a billing machine).

1

Mr W Smith
SB Arr 2/11–4/11
£40 + VAT
Own A/C

On arrival

	£ p
Tea	0.40
Telephone	1.20
Balance	
Lunch	4.80
Wine	1.75
Coffee	0.40
Balance	
Telephone	0.75
Accommodation +VAT	

Change the date

Day 2
Breakfast   1.75
Telephone   0.80
Balance

Mr Smith pays £30 in cash
towards his account.
Balance

Dinner   4.50
Beer   0.80
Coffee   0.40
Balance

Telephone 0.80
Accommodation + VAT
Balance

Day 3
Breakfast 1.75
Mr Smith checks out and pays the balance by Access

2

Ms J SMYTHE
Single room 115
£30 + VAT

Arrival 2nd July for 1 night
Ms J Smythe gives a deposit of £50 on arrival

Tea   25p
Telephone 40p

Balance
Lunch £2.50
Wine   £1.50

Balance
Tea   25p
Telephone £1.50

Balance
Theatre tickets £20.00
Bar   £ 5.00

Balance
Room
VAT

*She pays the
difference by
cheque and leaves
the hotel*

3

Three people
are in residence
when you come
on duty at 3pm:
Mr P Johnson   SB   £30 + VAT   B/f 29.80   Room 301

Ms W Jackson   S   £20 + VAT   B/f 15.40   Room 112
Mr S Thompson   S   £20 + VAT   B/f 24.50   Room 109

THE FOLLOWING CHARGES ARE TO BE POSTED:

- Room 112   telephone 25p and 75p
- Room 112 and 109   afternoon tea @ 60p per person

- Room 301   dinner £4.90, wine £2.00, coffee 40p
- Room 109   dinner £6.00, beverages £1.20, coffee 40p
- Room 112   dinner £7.80, beers £1.20, wine £3.50, coffee 80p

- Room 301 pays £20.00 towards his account by cheque.

Post accommodation to all rooms

Take an 'X' reading   to check before taking the 'Z' reading.

4

Zapata Oil, 3 Old Bond Street, London W1
The above company has a function in the hotel on 15th November
Prepare the invoice from the following information.

Hire of the Princess Function Room £100.00 plus VAT
Lunch £7.50 per person plus VAT
Wine £2.00 per person plus VAT
Coffee 20p per person plus VAT
Service charge of 10% on net price of food and wine
50 people attend the conference
Account to be sent to the company for settlement.

5   Day 1

Commencing 1st March

102	103	104	105
SB £40 + VAT	TB £60 + VAT	TB £60 + VAT	SB £40 + VAT
Ms P Knight	Mr/s J Peters	Mr/s D Jarvis	Mr P Andrews
28/2–4/3	1/3–2/3	1/3–5/3	1/3–4/3
£48.20 (B/F)			

£100 is paid on arrival by Room 103 in cash
Telephone calls (104 £1.20) (105 £1.75)
Morning coffee (102 80p) (103 £1.60)
Lunch: 105   £18.00, wine £2.20, coffee 50p, liqueurs £2.20
       102   £4.80, minerals 75p
Telephone calls: 105 80p, 103   £1.25

Room service: 104 £4.80
Theatre tickets were bought for Room 102 £8.75 from an outside agency.
Room 105 was overcharged by £1.05 at lunch time. Make the adjustment.
Room 102 leaves unexpectedly. As it is 1700 × 50% of the room charge
is made. Ms Knight pays by Access card.
Dinner for 104 £28.00, wine £5.75, coffee £1.00.
Chocolate and sandwiches for room 105 £1.75.

Post the accommodation charges for the next day and take a trial balance.
Then take a Z reading. What is the C/F figure for the next day?

6 Day 2 2nd March

103	104	105
TB £60 + VAT	TB £60 + VAT	SB £40 + VAT
Mr/s J Peters	Mr/s D Jarvis	Mr P Andrews
1/3–2/3	1/3–5/3	1/3–4/3

Introduce a cash card
Breakfast: **103** £2.20, **104** £2.20, **105** £1.10
            Restaurant cash takings £47.20
Telephone calls: **104** 75p, **103** £1.25
Mr/s Peters depart and pay cash.
Mr W Brown arrives and is allocated Room **102** @ £40 + VAT. He is
staying two nights and will be paying his own account.
Coffee: **102** 80p, **104** £1.60
Flowers bought from outside for Room **104** @ £4.80
Lunch: cash takings £108.50 Drink £82.40
        104 £10.80, wine £4.50, coffee £1.00
        102 £7.50, wine £2.80, coffee 50p
Cigarettes: **105** £2.80
Room service: **102** £1.80
Room **104** was undercharged for dinner on 1/3. An extra charge of
£2.20 for liqueurs has to be entered onto his bill.
Dinner: **102** £38.50, wine £18.70, coffee £2.00
105 £9.50, minerals £1.20
Cash takings £309.40 Wine £127.00
Room service: **104** £2.50, coffee £1.00
Telephone: **102** 95p, **105** £2.75

Post the accommodation for the next day, take X reading. Check the cash
figure. How much cash have you taken on your shift? Take Z reading and
C/F for next day.

7 Day 3   3rd March

102	104	105
Mr W Brown	M/M D Jarvis	Mr P Andrews
SB £40 + VAT	TB £60 + VAT	SB £40 + VAT
2/3–4/3	1/3–5/3	1/3–4/3

Introduce a cash card and a ledger card

Breakfast: **102**   75p, **104**   £1.50, Restaurant cash £32.40

Telephone calls: **105**   £1.60

Coffee: **104**   £1.60

Lunch: cash takings £67.20, wine £31.50, coffee £7.50,
   102   £8.40, minerals, 85p, **105**   £4.90, wine £2.50.

Visitors paid out for theatre tickets for Room **104**   £9.75 (A service charge of £1.00 was added to the original cost.)

Dinner: **102**   £8.90, wine £5.60, coffee 60p,
   105   £10.50, wine £4.90, coffee £1.20.
   Cash takings £280.49, wine £123.00, coffee £32.40.

There was a function held in the Waverley Suite and the following costs were incurred. Make up an invoice for payment.

Room hire £80.00 + VAT

45 Dinners   @ £7.50   per head + VAT

Wine   @ £2.00   per head + VAT

Coffee   @ 30p   per head + VAT

The account was forwarded to John Brown and Co. Ltd, 29 Down Street, London W1 for the attention of Mr W Saunders

Work out the carried forward by using an X reading. Then take a Z reading.

6 Cashiers and methods of payment

1 Terminology

You are the youngest member of your team, and have recently completed your course successfully. Your hotel is co-operating with the local college and is accepting a student on industrial release. As she is close to your age you have been appointed her mentor, and the head cashier has asked you to ensure she is familiar with the terminology in general use.

Task

Explain in the clearest way possible what is meant by the following terms and when they are likely to be encountered.
- the payee
- the drawer
- the drawee
- bank sorting code
- current account
- deposit account
- special clearance
- merchant
- merchant number
- cardholder
- sales voucher
- imprinter
- authorisation code
- stop list

45 minutes

2   Foreign currency exchange

```
┌─────────────────────────────────────────────────────┐
│ FOREIGN CURRENCY EXCHANGE                             │
│ ───────────────────────────────────────────          │
│                              0069                     │
│                                                       │
│  Room No:................      Name:.................. │
│                                                       │
│     Currency:........................................ │
│                                                       │
│     Amount:.......................................... │
│                                                       │
│     Sterling Equivalent: ............................ │
│                                                       │
│                                                       │
│  Date:..................      Cashier:............... │
│  LAYSTON                                              │
└─────────────────────────────────────────────────────┘
```

Mr McKie (room 102) presents 500 Belgian Francs to exchange for sterling. The exchange rate is currently BF64.65 – £1.

Tasks

1 Complete the exchange slip.
2 Explain why a hotel would require the cashier to complete a foreign exchange slip in duplicate.
3 Who would receive:
   (a) the top copy
   (b) the duplicate?

20 minutes

3   Speedy check out

When on late duty as cashier you are handed this slip by a guest who is leaving tomorrow morning.

NAME _____ MR/MRS

ROOM NUMBER _____

CARD

ACCESS	VISA	AM EX	DINERS	THF	JCB
☐	☐	☐	☐	☐	☐

NUMBER ☐☐☐☐☐☐☐☐☐☐☐☐☐☐☐

EXPIRY DATE _____

DATE OF DEPARTURE: _____
Address to which account copy should be sent:
(CAPITALS)

....

....

....

If settling your account by Credit Card we invite you to take advantage of our rapid check-out facility. On departure simply complete the details and place this card with your key at the key drop on the reception desk.

A copy of your account together with your credit card voucher will be sent to you on your day of departure.

Our receptionists will be delighted to make any onward hotel reservations within Trusthouse Forte in the U.K. or worldwide.

I hereby agree that my credit card should be used to settle the full amount of my bill and sent to the stated address.

Signature ....

Tasks

1 Explain what the slip is for.
2 Highlight the main advantages:
   (a) to the hotel
   (b) to the guest.
3 How could this system be adapted and more widely used?

30 minutes

4 Foreign currency

The hotel has decided to charge two per cent on the bank rate in order to cover any fluctuations in currency.

Task

Complete this table which is to be used for display purposes at the cashier's desk.

Currency	Bank Exchange Rate	Commission	Hotel Display Rate
French Francs	9.79		
German Marks	2.93		
Italian Lire	2125.00		
Spanish Pesetas	201.00		
US Dollars	1.50		
Swiss Francs	2.44		

30 minutes

5 Issue of floats

You are the front office cashier and you have been asked to give banqueting a float for the cash bar for the Women's Circle Annual Dinner and Dance. A sum of £100 has been authorised.

Tasks

1 List the procedure you would follow when handing it over, and itemise how you would make up the float.
After the function the barman pays in as follows:

Notes	Coins	Cheques	Others
1 × £50	28 × 1	J Smith – 11.75	Travellers' cheque –
2 × £20	31 × 50p	F Brown – £14.40	£20
21 × £5	210 × 10p	W Willis – £8.90	Access Voucher
	90 × 5p	P Jones – £16.50	£25.50
	82p copper	L Lovitt – £15.20	American Express
		M S Murphy –	£18.50
		£18.95	

2 The Manager has asked you to make a special banking for the function. Compile and complete a cashier's report with all the necessary information.

45 minutes

6 Petty cash

Situation A
As the cashier at the Windy Bay Hotel your tasks include the keeping of the petty cash.

Task

Enter the following transactions shown here in the Petty Cash Analysis sheet opposite, beginning with an Imprest of £50.

30 minutes

Petty Cash

Location .... **W/E** .... / .... / ....

VOUCHER No.	Total		Newspapers			Repairs	Gen. Exp.	Gen. Exp.	Cleaning	Advertising	DETAILS (Full Details on Voucher only)	VOUCHER No.
1												1
2												2
3												3
4												4
5												5
6												6
7												7
8												8
9												9
10												10
11												11
12												12
13												13
14												14
15												15
16												16
17												17
18												18
19												19
20												20
21												21
22												22
23												23
24												24
25												25
26												26
											Total Payments £	

Certified Correct .... **Manager** ....

Situation B

As cashier at the Majestic Hotel, your tasks include the keeping of the petty cash float and accounts. You have been attending an exhibition for a week, and payments made on petty cash have been paid out by the manager in your absence. He has kept a list of payments, but has not entered them into the analysis sheet.

Begin with an imprest of £50

Sep 1	stamps £8.80	
Sep 2	oasis for Housekeeper £9.20	
Sep 3	fresh ginger for chef £1.18	
Sep 3	taxi to HQ (self) £3.20	
Sep 4	repair of suitcase £5.50	
Sep 4	interviewing expenses (travelling) £18.20	
Sep 6	flowers £9.90	
Sep 8	stamps £9.60	
Sep 9	glue 80p	
Sep 10	A–Z for porters £5.65	

Task

Rule up a petty cash book with the appropriate headings and post the charges which have occurred in your absence, balancing and renewing the imprest as necessary.

45 minutes

7 Payment by cheque

The cheque below is offered to you as payment for an account. The client is leaving this morning, and you (as cashier) have three options open to you.

- accept the cheque and issue a receipt
- obtain a special clearance
- request an alternative method of payment.

Tasks

1 Explain the procedure involved in all three cases.
2 Explain the advantages and disadvantages of each option.
3 State which option you would favour.

30 minutes

8 Float reconciliation

You are a cashier at the Whitlock Hotel and your float is made up of the following:

- £10   £30.00
- £5   £30.00
- £1   £27.00

- 50p   £ 4.00
- silver   £ 8.25
- copper   £ 0.75

The following transactions took place while you were the cashier on duty.

Room	Account	Method of payment
201	92.40	£20 × 5
812	141.69	£20 × 7, £1 × 2
111	197.45	paid by Access
210	–	paid £50 on account in cash
Porter	–	gave a PCV for a taxi fare of £2.50
305	47.20	£50 × 1

An advance deposit of £60 (£10 × 6) for a room booked for the following night is given by Mr W Willis.

102	87.50	Paid by Visa
105	38.20	Paid by cheque
211	82.14	£10 × 8, £5 × 1
Chef	–	gives a PCV for a meat order for £25.50
202	62.80	£10 × 7
205	–	Exchanged $100 at the exchange rate of $1.12.
Restaurant	32.70	You accept payment in French Francs (500). The exchange rate is FF10. 38.

Tasks

1 List the exact contents of your float after the above transactions have taken place.

2 Prepare a statement to show what change and money you have given out.

3 What is the total amount of cash receipts that have been issued?

4 Complete the bank credit slips here showing what money will be paid into the bank, and then give the precise breakdown of your remaining £100 float.

Note:
Try to bank all the larger notes and to keep adequate money for using as change.
Assume, when banking, that the PCVs have been exchanged for cash.

1 hour

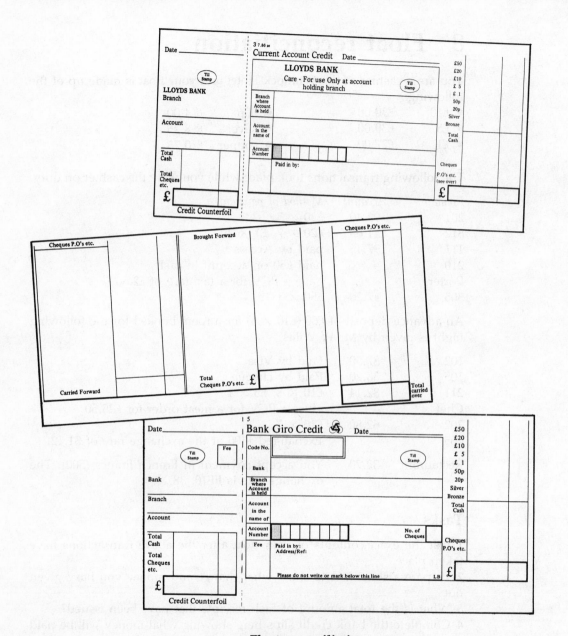

Float reconciliation

Foreign Credit Cards	Cheques	£50	£20	£10	£5	£1	50p	Silver	Bronze

9 Cash control

The following guests are departing today:

		Amount owing	Method of payment
102	Mr J Johnson	49.80	Midland Bank Cheque
210	Mr and Mrs R Smith	140.28	$ 200 (US)
212	Mr E Williams	8.50 cr	
213	Mr and Mrs O Brown	92.60	Eurocheque
214	Mr T Green	41.80	Cash £80
218	Mr S Grey	229.75	Diners Club
221	Mr and Mrs P Black	105.60	Ledger British Airways
249	Mr Smythe and Mr Hancock	170.59	American Express
302	Mr I Bates	78.60	Master Charge
304	Mr and Mrs P Stubbins	179.20	Travellers Cheques 2 × £100
306	Mr K Leedham	92.40	25 000 Pesetas

Notes

Methods of payment accepted (subject to usual rules and conditions)

American Express	£175.00	
Visa	£150.00	floor limits
Access	£100.00	
Diners Club	£200.00	

Travellers cheques (£1 commission is charged on each transaction)
Eurocheques
Bank cheques
Foreign currency (£1 commission is charged on each transaction)

Foreign currency accepted with exchange rates:

French Francs	10.20
US Dollars	1.35
Spanish Pesatas	221.00
Italian Lire	2350.00

Task

In each of the above cases list the procedure for accepting a particular method of payment, naming all the documentation that will be used, and where relevant what change to be given, and any authorisation required.

1½ hours

10 Bank paying-in

You are the general assistant in a small hotel (six rooms), and your tasks include almost everything.

The manager has left you to do the banking. You bank at Lloyds, but unfortunately not the local one, so you will have to go into Westgate-on-Sea. The name of the account is the Crislin Hotel, and your Account number is 0442310. The bank code number is 90-51-09.

The following need banking.

Cheques

Lloyds	(Gunner)	42.80
Midland	(Lewes)	53.76
Nat West	(Adams)	60.41
Barclays	(Collett)	15.25
Nat West	(Wilkins)	114.90
Lloyds	(Moss)	90.80
Coutts	(James)	78.87
Barclays	(Wells)	59.70

Cash £10 × 18, £20 × 20, £50 × 15, £5 × 21
£1 × 21, 50p × 17, 10p × 80, 20p × 20, 2p × 55

Task

Complete the appropriate paying-in slips.

45 minutes

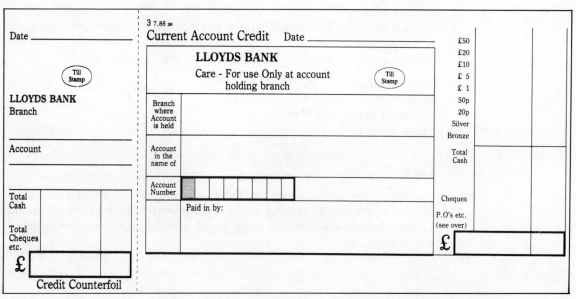

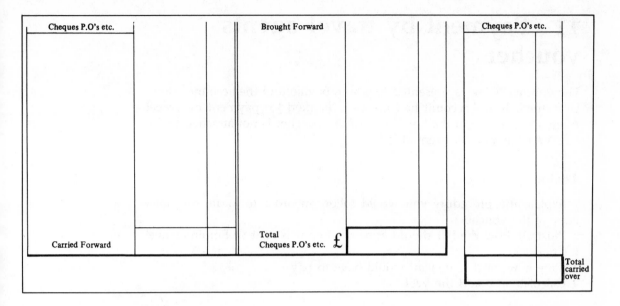

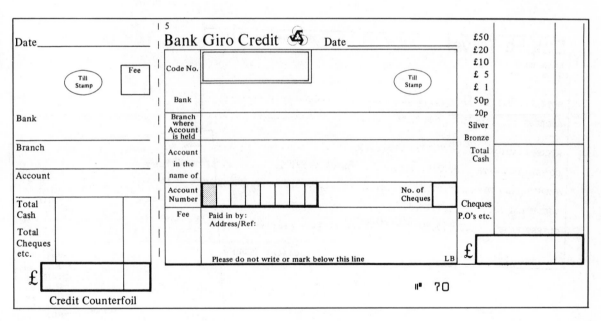

11 Payment by travel agents voucher

The voucher below is presented to you in payment of the account below. Unfortunately the account had not been marked for payment by Travel Agents Voucher, and the hotel copy of the voucher is not attached to the bill. VAT has not yet been added.

Tasks

1 Explain the procedure you would follow in order to locate the hotel copy of the voucher.
2 Illustrate how the bill should appear when it is sent to Empire Travel for payment.
3 State how much the client would have to pay.
4 Calculate and add the VAT.

1 hour

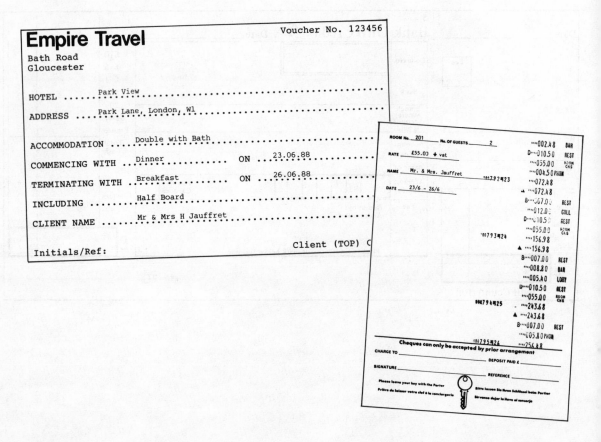

12   Methods of payment

You are the duty cashier and a guest (leaving today) whose bill totals £98.20, offers you a choice of methods of payment as illustrated below.

Tasks

1 Which of the methods would be:
(a) most secure,
(b) most beneficial financially to the hotel?
2 Describe the procedure to follow when accepting *any* of these methods of payment.

1 hour

× 10

× 8 @ $1.50

× 10 @ 10.32

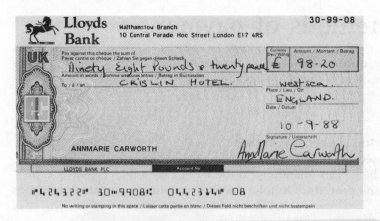

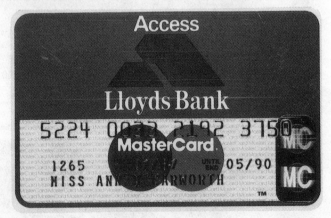

13   Cashier's summary

The hotel where you work is small, and has a manual system. All receipts are issued by hand from a duplicate book, and at the end of the shift the cashier summarises the payments taken and deposits the takings with the Manager, who does the main banking.

During the day the following occur:

10	Miss Wynne	£50.30	Cash	Receipt No. 0214
40	Mr Pampellone	£60.49	Access	
32	Mr/s Joyce	£90.30	Mastercharge	
21	Miss Miller	£14.90	Cash	
13	Mr Noonan	£80.80	American Express	
16	Mr/s Charles	£115.08	SL folio S/73 Charles Bros. Ltd	
19	Mr/s Slapper	£211.76	Visa	
38	Miss Moumouri	£79.00	Eurocheque	
49	Mr Martins	£76.89	Cheque	
11	Miss Wastell	£10.23	Cheque	
17	Mr MacGregor	£25.45	Time Tours Ltd (Voucher No 4193)	
12	Miss Dilaver	£78.90	Cheque	
Cocktail bar		£91.99	Cash	
09	Mr Constantino	£63.55	Cheque	
22	Miss Melambianaki	£67.70	Visa	
23	Miss White	£76.43	Diners Club	

At the end of the shift your cash is broken down as follows:

£50 × 1
£20 × 2
£10 × 5
£5 × 2
£1 × 5
50p × 4
Copper – 19p

Prior to banking the following occurs:

Mrs Pampellone changes FF 200 – the exchange rate is FF 9.28
Miss Bell exchanges 50 Deutsch Marks – rate is DM 3.89

Tasks

1 Make payments for foreign exchange from day's cash takings.
2 Enter all transactions on Cash Sheet (following through the sequence of receipt numbers).
3 Complete the Paying In Summary.

1½ hours

CASH SHEET

NAME .... Date ....

Receipt No.	Room No.	Name	Total	RECEIPTS Cash	Sales Ledger	Remarks

TOTAL RECEIPTS

PAYING IN SUMMARY

Dept. ....

Name .... Shift ....

Signature .... Date ....

Foreign Cheques	
U.K. Cheques	
Irish Cheques	
Travellers Cheques U.S.	
Travellers Cheques U.K.	
Travellers Cheques Others	
$ U.S.	
Francs French	
D.M.	
Guilders	
Francs Belgian	
Lira	
Francs Swiss	
Other Foreign Currency	
£50	
£20	
£10	
£ 5	
£ 1	
50p	
10p	
5p	
2p	
1p	
Total Banked	
Over/Under Banked	
Actual Cash Business	

7   Tours and groups

1 Delayed flights

Your hotel has just been contacted by the Air Scribe airline who are grounded at the nearby airport because of an industrial dispute. Flights are leaving from the East Midlands (20 miles away) but there is a backlog and the Air Scribe flight is on a waiting list: they will not be able to leave until tomorrow. Part of the delayed flight, who are students, are booked into your hotel for tonight. An Air Scribe coach will arrive for them at 0830 tomorrow, to take them to the airport. The airline is paying for dinner, bed and breakfast and has reserved the last of your rooms; 7 twins and 10 triples, for 42 students and 2 teachers. When they arrive you discover that the teachers are 1 male and 1 female and the students are 15 females and 27 males.

Tasks

1 Explain how you will allocate the rooms to the group.
2 List the other departments in the hotel who will need to be advised of the delayed flight.
3 Design a brief handout to be given to the students on arrial to ensure that they behave in a considerate manner and do not inconvenience other guests.

30 minutes

2 Group bookings

You are the Reservation Manager of a 270 room four star London Hotel. The hotel has 90 twins, 90 doubles and 90 singles. Rack rates are £90 twin and double, £75 single per night, including service and VAT. The restaurant seats 100, the coffee shop 120 and there is a banqueting room available which seats 300 for lunches or dinners.

   Meal rates for tours above 50 people are:
Continental breakfast   £4.50
English breakfast   £6.00
Lunch   £9.00
Dinner   £10.50

You receive the following telex from a New York travel agent on February 10th:

Please quote availability and charges for Ameritour 88 Group
Arrival Sunday 13/4/88 1430 from Luton Airport
Depart Wednesday 16/4/88 (3 nights) Heathrow at 1000
70 Twins and 50 Singles
Dinner and breakfast required

Tasks

1 State the steps you would take from receipt of this telex, through to guests and courier going to their rooms on arrival six weeks later.
2 Describe the Reservation Manager's procedure in dealing with tour bookings.
3 Describe the Front Office Manager's procedure in dealing with tour bookings.

45 minutes

3 Tour bill

JAC Travel
24 Compton Street
London W1
Telephone: 01 373 2154
Have booked accommodation at your hotel for 70 people (+ driver of coach) for arrival 2nd September for three nights. They require 6 Singles with bath, 10 Twins with bath, 15 Doubles with bath and 5 Triples with bath.
Estimated time of arrival 1430.
One single room complimentary for the driver has been agreed, but his meals are to be charged at the same rate as the rest of the group.
English breakfast is required each morning at £1.50 pp + VAT.
Lunch is required on the 4th September at £4.50 pp + VAT.
Room, breakfast and lunch account to the agent for settlement.
Extra's account for each guest to be opened.
Rates quoted:
Single Bath £15 + VAT
Twin bath £22 + VAT
Double bath £22 + VAT
Triple bath £29 + VAT

Tasks

1 Enter the above information onto a group breakdown sheet.
2 Prepare a pro-forma invoice which is to be sent to the Travel Agent before arrival for approval. A deposit of 20% of the total account has to be sent to secure the booking.

45 minutes

```
                    Group breakdown
Group name                              No of pax
Agent name

Agent address


Telephone

Accommodation
 SB
 TB
 DB
TRB
                                   Accomm
                                   VAT _____
                                   Total
Arrival                  Departure

Arrival details

Complimentary

Meals


Account instructions



Allocations
```

4 Conference sales

Refer to the Hotel A in Appendix 1.
The Regent Room can accommodate the following number of people:

- Boardroom style   45
- Theatre   90
- School room   55
- Cocktails   160
- Traditional banquet   90
- Dinner dance   80

Other information:
Floor area: 140 m
Access from street or through lobby
Air conditioning
Dance floor
Adjoining bar 5 metres × 10 metres

Tasks

1 From the information provided design a conference check list which can be used both to assist the organiser, and as a sales aid for hotel staff.
2 As a group compile a brochure for this facility.

45 minutes

5 Banqueting sales

The fixed costs of operating your banqueting room in the hotel are £80 per day. You have a potential customer who has selected a menu for one hundred people. She has booked the room before, and is likely to do so again, but she does not want to pay more than £3.75 per person. The banquet is at very short notice, five days away. Your hotel is very quiet at the moment and it is most unlikely you will have the chance to let the room to anyone else.

The menu chosen by the customer has a food cost of £2.40 per person, a variable wage cost of 70p per person, and other variable costs of 10p.

Tasks

1 Calculate the total price per person if this banquet were booked.
2 What should the total selling price be if 15% income on revenue is wanted.
3 What should the price per person be to gain 15% net income on revenue.
4 Bearing in mind the price your customer wants to pay, the price it will cost the hotel and any other factors which may be relevant, briefly outline your decision, stating why you have accepted or rejected this booking.

45 minutes

6 Wedding sales

Two young people approach you one weekend when there is no-one on duty in the banqueting department, and the Food and Beverage Manager is already engaged. They are getting married next year, and recently attended a wedding in your hotel and were very impressed.

Referring to the information given in Assignment 5 on *conference sales*, carry out the following tasks.

Tasks

1 Explain how you would handle this situation, with a view to making a provisional sale.
2 List other advice and services which your hotel could supply to ensure that the wedding reception is a success.

45 minutes

8 Sources of business and selling

1   Sales promotion

You have been working for two years as a Chef de Brigade in a large West End Hotel, when the Manager approaches you and asks you to join the Sales Department. This is a promotion for you, with an increase in salary and the privilege of being able to go out of the hotel, and to entertain clients.

The hotel is trying to generate more business from local commercial companies for both bedroom occupancy and for banqueting usage.

Tasks

1 Describe what equipment or information you would need initially, and then explain exactly how you would go about trying to generate more business. Some ideas may be expensive but if you can justify the expense they may be accepted.

30 minutes

Use any one of the three hotels in the Appendix, or the hotel where you spent your industrial release, or any other establishment well-known to you.

2 Prepare an intensive selling campaign to be carried out by the reception staff during the quiet months of January and February.

Each member of the team must fulfil some function, and the activity should have a measurable result.

30 minutes

2 Sales leads

The general manager is intent upon maximising sales and has sent you the information shown below and over leaf.

He sees them all as sales leads, and wants you to follow each one up, either by telephone, letter or any other way which you think would be effective in producing increased sales.

Tasks

1 Show how you would deal with each of the following:

Extract from local paper – Week ending 6th March

> Mr and Mrs White are pleased to announce the engagement of their daughter, Sarah Verity, to Russell Joseph. The wedding is planned for early July.

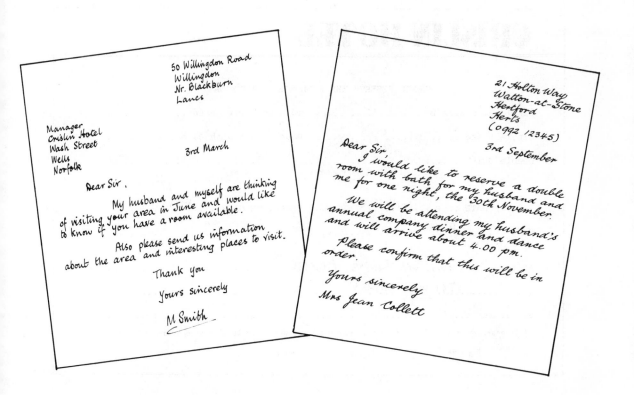

50 Willingdon Road
Willingdon
Nr. Blackburn
Lancs

Manager,
Crislin Hotel
Wash Street
Wells
Norfolk

3rd March

Dear Sir,

My husband and myself are thinking of visiting your area in June and would like to know if you have a room available.

Also please send us information about the area and interesting places to visit.

Thank you

Yours sincerely

M Smith

21 Holton Way
Watton-at-Stone
Hertford
Herts
(0992 12345)

3rd September

Dear Sir,

I would like to reserve a double room with bath for my husband and me for one night, the 30th November.

We will be attending my husband's annual company dinner and dance and will arrive about 4.00 pm.

Please confirm that this will be in order.

Yours sincerely
Mrs Jean Collett

Room No   Date

Dear Guest,

I understand you intend leaving tomorrow. If so, it would greatly assist us
in preparing for incoming guests if you could vacate your room by 12 noon.
If, however, you would like to stay on, please let the Reception Desk know
at once. They will do all they can to look after you.

May I, on behalf of everyone on the hotel staff, thank you for coming to
stay with us, and hope that you will come again soon. If you would like
to comment on our service, please do so on the reverse side of this card.
Compliments will be passed on; criticisms and suggestions are welcome and
will receive my personal attention.

We would be most happy to welcome you again and attached you'll find a room
reservation card for your use. Just simply fill in your requirements, detach,
and leave the card with Reception. Or if you prefer you can mail it - no
stamp is required.

Yours sincerely

General Manager   PTO

CRISLIN HOTEL

PLEASE RESERVE THE FOLLOWING ACCOMMODATION

Single Room with Bath ...... ✓ .... Twin Room with Bath ....

Single Suite .... Double Room with Bath ....

Double Suite ....
                    (Please insert number of rooms required)

ARRIVAL DATE .... 2nd Mar .... DEPARTURE DATE .... 5th Mar .....

Name .... MISS E PAM ....
(Block letters)

Address ..... 28 HIGH GROVE AVENUE ....

.... DELTON WARKS ....

Signature ..... E Pam ....

This reservation will be confirmed to you and the booking should then be considered definite.

* I like the room I had this week.

114

Reason for purchases

The photograph is of one of your TB. The room is located on the eighth of nine floors and overlooks the nearby downs. It is, however, near the lift and quite noisy. It is the last room you have available tonight and is in the top price range for the TBs.

A customer is standing at the desk, but is obviously hesitant about taking the room.

2 Explain how you would endeavour to sell the room, using the basic reasons which influence customer purchase, as listed below
- need
- comfort
- desire
- pride
- pleasure
- fear
- fashion

45 minutes

3 Marketing information

Four different registration cards currently in use in hotels are reproduced here.

A (Front)

For Office Use	
Room Number	
Arrival Date	
No. of Nights	
Rate (£)	
Remarks	

SURNAME

FORENAME

HOME ADDRESS

Mkt. Type   Persons

Account to be settled by:
Cash   ☐   Voucher   ☐
Credit Card   ☐

Nationality

Car Reg. Number

OVERSEAS VISITORS
Passport Number

Signature

Issued at

Receptionist

Next destination

Dept. Date   Name

(Back)

RETURN VISITS

Date	Room	Nts	Rate

THE CAPITAL HOTEL
Knightsbridge

NAME: ....

ADDRESS: ....

....

....

VEHICLE REG. NO. : ....

	ARRIVAL DATE: ....
NATIONALITY:	DEPARTURE DATE: ....
PASSPORT NO. :	NO. OF NIGHTS: ....
ISSUED AT:	NO. OF GUESTS: .... ROOM No. : ....
DESTINATION:	ROOM RATE: ....
....	

SIGNATURE ....

B

WELCOME
to the

KENSINGTON CLOSE HOTEL

Room No.

GUEST REGISTRATION CARD

Please Print your Personal Details.

Mr/Mrs/Miss		Nationality	
Home Address		Passport No.	
Town City	State County	Place of Issue	
Country	Postal Zipcode	Destination	

Company/Travel Agent

Address

Signature

I agree that my liability for this account is not waived and agree to be held personally liable in the event that the indicated person, company or association fails to pay for any part or the full amount of these charges.

Account Settlement

- ☐ Cash
- ☐ Visa
- ☐ Access
- ☐ Amexco
- ☐ Diners
- ☐ THF
- ☐ Com. A/c.
- ☐ Airline/TVL Voucher No.

FOR OFFICE USE ONLY

Guest Tarriff	Arrival Date	Departure Date	Arrival Time	No. of Rooms	Room Type	Room Rate Package	
No. Guests	No. of Children	Adv Deposit	Voucher No.	Receptionist	Supervisor	Origin Code	

C

D (Front)

EROS HOTEL

69 SHAFTESBURY AVENUE LONDON W1V 8EX
Telephone: 01-734 8781 Telex: 268564 EROSHT

Length
of stay

Number
in party

CASH ☐
CHEQUE ☐
A/C ☐
CREDIT CARD ☐
VOUCHER ☐

MR.
MRS.
MISS _____
Please Print in Full

Address _____

_____

_____ City _____

Country _____ Nationality _____
Car Registration
Number _____

Passport No.
(Aliens only) _____
Where
Issued _____

Next Destination
(Give address) _____

ACCOUNT CUSTOMERS ONLY — If the Company for whatever reason fails to honour this account, I under-take to be held personally liable for the full payment thereof.

Receptionist | Signature

(Back)

	a.m.			
Arr.	p.m.	Name		Room No.
Dep		Rate	Acc Requested	Length of stay

				Deposit
Letter				
Telex		Date booking made	Receptionist	Receipt No.
Cable				
Counter		Booked by		Voucher No.
Telephone				
Chance		A/c to.		
Regular				
T.B.C.				
N.T.B.C.				
Remarks				

Tasks

1 List the marketing information which can be extracted from each card.
2 For each card list the information which is legally required, against that which is required by the hotel.
3 Compile a list of information which you think would be useful to include upon a registration card.

30 minutes

4   Referral bookings

You are employed in Hotel C in the Appendix. The hotel is privately owned, but next year will be joining a marketing consortium to gain full advantage of referral bookings.

Task

Using the hotel in the appendix as a guide, and providing yourself with additional information as necessary, compile a sales check list for the other members of the consortium.

30 minutes

5 Questionnaires

Refer to Hotel A in the Appendix. A guest questionnaire is to be placed in all rooms to enable guests to comment on services and facilities. It has been decided to keep it simple, e.g. ticking off boxes or marking grades A to E, rather than asking the guests to write too much, which may discourage them from completing it. However it would be a good idea to leave space for comments or suggestions.

Tasks

1 Accepting that the hotel product consists of accommodation, food, drink and service, design a questionnaire which enables guests to comment upon all these services and facilities.
2 Explain the procedure for handling a completed guest questionnaire when a complaint has been made.
3 Draft a suitable letter to be sent to the dissatisfied customer in order to try and retain their business.

1 hour

6 Shops and concessions

Many of the hotels in your town have granted concessionary leases to shops and agents, providing an additional service to customers. There is an area near reception which is under-utilised, and the management have decided to convert it into two shops, with a small area for the display of leaflets and brochures. The first shop is to be stocked and run by the hotel, the second section is to be let.

Tasks

1 List the items which should be on sale, giving reasons for your choice.
2 To whom would you approach regarding the rental of the second shop? State why.
3 List the leaflets and brochures which you would stock for use by both business/tourist sectors of the market.

1½ hours (including research)

7 Selling up

A customer arrives at the reception asking to see one of your double rooms. She is planning to bring her husband for a long week-end as a surprise to celebrate their 25th wedding anniversary.

The customer is slightly known to you as the accountant of a successful company in the town.

The hotel is fairly quiet and you are able to leave your colleague at the desk to show the potential client a room.

Task

Describe in your own words the selling process in this case, including:
- the opening of the sale
- the developing of the sale
- the closing of the sale.

20 minutes

8 Regular guests

Your hotel has only been open for 18 months but an analysis of your sales figures show that an increasing percentage of business is repeat bookings. The Manager is concerned to ensure that this sector of the market is developed and has suggested that the Front Office maintain a file of regular guests.

Tasks

1 Design a suitable form for a Guest History Card.
2 Suggest ways in which repeat business can be encouraged.
3 List some incentives which could be offered to regular clients to ensure their continued loyalty.

15 minutes

9   Activity weekends

Your hotel, situated in the Lake District, has decided to market special activity weekends as follows:
- fishing
- golf
- rambling
- artists
- gourmet

Tasks

1 Design a short advertisement to place in a newspaper/magazine.
2 Suggest a suitable specialist magazine for each of the activity weekends to be advertised in.
3 Cost the advert for both a one–off and a series advertisement in:
   (a) a national newspaper (supplement)
   (b) a specialist magazine.
4 Design a suitable package, commencing with dinner on Friday and ending with afternoon tea on Sunday for at least one of the activity weekends.

1 hour

10   Research

You are employed in a 30 bedroom hotel in Wales.

Many establishments of similar size in the area have devised an 'all-in' package, often with a theme. Your hotel has no swimming pool and is not close to a golf course, but it is only 20 miles from Hereford and in a charming setting, close to the foothills of the Black Mountains and in grounds of its own.

The hotel has been modernised, and all rooms except five include an en-suite facility. There are two bars, both of which are open to the public, a restaurant and a games room in the basement. The hotel is not 'olde worlde', but the public rooms are large and comfortable, and include a small banqueting/conference room which is popular for weddings.

Task

1 Work in groups of 3–4 for this assignment. Begin a file of information about the area and the particular locality of the hotel. Allocate tasks, i.e.:
- writing to the Tourist Board of Wales.
- visiting a local travel agent.
- checking in the library, etc.

2 From the information/statistics/brochures you have collected, try to identify a potential market for special breaks, identifying customer needs and wants.

3 Use the information to suggest a possible 'time-table' for a special break. List the speciality themes which the geographical location suggests may be possible.

4 Plan and produce a brochure for your hotel, furnishing yourself with information as necessary. Base your prices on similar brochures.

5 Explain ways in which a suitable mailing list could be compiled.

6 Suggest how you would decide upon the distribution of the brochure, and where advertisements could be placed, giving reasons and approximate costs.

4–6 weeks

Note: If you are taking C & G 720, this presents an excellent opportunity to demonstrate your ability to work together in groups.

11 Planning a promotion

As part of a plan to increase the number of Canadian visitors to London, four leading hotels are offering a five day complimentary stay for the agents to enable them to sample the hotel, use the facilities and visit some of the tourist attractions.

Each hotel can accommodate ten agents, spreading the cost, and will arrange for one function and a tour of the hotel for all travel agents to attend, along with the Sales Managers and General Managers from the other hotels. Your hotel is anxious to make a good impression on the travel agents as well as the competition.

The Crown Hotel has submitted a programme for day 2 as follows:

9.30–10.00	Breakfast for ten agents
10.15–12.00	Coach departs to St Katherine's Dock for sight-seeing, picking up other agents en route.
12.00–12.15	Coach to Charing Cross Pier, lunch on board 'The Primrose' and a visit to the Thames Barrier.
4.00– 4.30	Coach back to hotels.
6.00– 7.30	Coach to House of Commons for cocktail party.
7.30– 8.00	Transport to the Crown Hotel.
8.00– 8.30	Tour of facilities, rooms, restaurant and bars in groups of ten.
8.30–12.00	Dinner in the Royal Suite (traditional cockney fayre), live entertainment by Chas and Dave and by traditional Pearly Kings and Queens.
12.00	Transport back to respective hotels.

Donations: Ladies' silk scarves from Blooms Department Store.
After-dinner cigars by Wilsted Tobacco Company
Table decorations by Candrels Florist.

Tasks

Use Hotel A in the Appendix (p. 168).
1 Suggest the most suitable room allocation for this group, giving reasons why.
2 List any additional items which could be placed in the rooms to make the guests feel more welcome.
3 Plan an itinerary for day 3 with one of the main meals at your hotel. Ensure that at least one of the main sight-seeing attractions will be visited. (Their enjoyment of this day depends on your organisation and planning.)
4 Write a welcome letter for distribution to each agent to accompany your brochure pack.

5 Draft at least one other letter to a local company requesting help regarding complimentary gifts or entertainment, stating why this would be of value to them.

6 The General Manager wishes to give each of the ten agents a small souvenir of the hotel on their departure. Suggest appropriate gifts and state why they would be suitable, to enable him to make his choice.

Since the sales promotion in March 1985 the following statistics have been recorded.

Room nights

	Jan	Feb	Mar	Apr	May	Jun	Jul	Aug	Sep	Oct	Nov	Dec
1984	4	3	9	12	15	19	23	28	25	17	9	7
1985	6	6	15	27	31	39	48	47	52	42	32	25
1986	21	19	23	28	33	36	49	48	49	40	35	30
1987	20	25	27	31	42	39	19	23	12	9	7	8

7 Plot the room nights onto a graph for visual display in the Sales Office, clearly showing the difference between the years.

8 Suggest ways in which the hotel can follow up the contacts made, to encourage continued business from the Canadian Travel Agents.

3 hours

9 Statistics and calculations

1 Guest statistics

Previous records show that for the first week in August, over the last three years, the following reservation pattern has emerged in a 300 bedroom hotel which consists of 85 SB and 215 TB.

		1985		1986		1987
Non arrivals	TB	5	TB	5	TB	6
	SB	2	SB	3	SB	5
Cancellations	TB	12	TB	11	TB	12
	SB	6	SB	6	SB	9
Early departures	TB	18	TB	17	TB	19
	SB	5	SB	7	SB	6

Tasks

1 On the basis of the above information, calculate a suitable percentage for overbooking for each type of accommodation.
2 Suggest ways in which non arrivals could be minimised.
3 Develop a plan of action for investigation into why the hotel has early departures.

45 minutes

2 Cocktail party costings

The Sales Department of the Westleigh Hotel are hosting a cocktail party for 85 travel agents from America, who are in town on a familiarisation trip. The General Manager has decided that a cocktail party is a good idea, but wants the Sales Manager to use money from the sales budget to cover the cost.

Tariff
Banqueting room £110.00
Canapes £15.00 per person
Drinks costed at £12.50 per person
The Food and Beverage Manager has agreed to give the Sales Department a discount of 15%.

There will be 11 members of staff attending the cocktail party and the General Manager will bear the cost of their attendance.

Tasks

1 Work out the total cost to be charged to the Sales Department. Show charges separately.
2 List the charges to be incurred by the General Manager.
3 List the members of staff who would be invited, giving your reasons.
4 Rule up a bill suitable for presentation to the General Manager and the Sales Manager, and enter the charges as appropriate.

45 minutes

3 Weekly analysis

The following table shows the accommodation let over a one week period in your hotel.

	TB	DB	SB
Monday	18	9	8
Tuesday	17	11	8
Wednesday	17	14	7
Thursday	15	12	9
Friday	9	9	6
Saturday	8	8	4
Sunday	4	5	3

Task

Enter the above figures onto the Weekly Report sheet and find the total
- (a) rooms
- (b) sleepers
- (c) income.

45 minutes

Weekly Report

Rooms	Sleepers	Income	Rate
TB 20			£75.00
DB 15			£70.00
SB 10			£62.00
Total			

	Rooms			Total	Slps	Income			Total income
	DB	TB	SB			TB	SB	DB	
Monday									
Tuesday									
Wednesday									
Thursday									
Friday									
Saturday									
Sunday									
Total									

4 Foreign currency conversion

Tasks – using current exchange rates

1 A guest's account is £78.20. How much is that in:
  (a) US dollars
  (b) French Francs
  (c) Deutsch Marks
  (d) Lira
  (e) Pesetas
  (f) Guilders
  (g) Australian Dollars
  (h) Krone
  (i) Swiss Francs

2 If a guest wants to exchange currency how much sterling would you give for:

(a)	French Francs	2000.00
(b)	US Dollars	150.00
(c)	Deutsch Marks	200.00
(d)	Lire	80 000.00
(e)	Pesetas	11 000.00
(f)	Guilders	490.00
(g)	Australian Dollars	2000.00
(h)	Krone	5000.00
(i)	Swiss Francs	10.00

3 If a guest's account is £62.20 how much change in sterling would you give if you received the following amounts?

(a)	US Dollars	100.00
(b)	French Francs	800.00

4 How much sterling would you give for the following amounts of foreign currency?

(a)	Deutsch Marks	82.00
(b)	Lire	9400.00
(c)	Pesetas	450.00
(d)	Guilders	15.00
(e)	Australian Dollars	50.00
(f)	Krone	120.00
(g)	Swiss Francs	75.00

5 If an account was £125.60 how much change in sterling would you give for the following amounts of foreign currency?

(a) US Dollars 200.00
(b) French Francs 1500.00
(c) Deutsch Marks 500.00
(d) Lire 400 000.00
(e) Pesetas 28 000.00
(f) Guilders 600.00
(g) Australian Dollars 200.00
(h) Krone 1500.00
(i) Swiss Francs 500.00

30 minutes

5 Room statistics

The Farrow Lodge Hotel has 50 rooms: 40 doubles and 10 singles. The rates are
- £20 per person per night
- £25 charge for double with single occupancy per night

The following figures relate to one week in September:

	M	T	W	Th	F	S	S
Singles	4	5	8	10	8	4	2
Doubles	23	25	40	40	21	18	7
Doubles S/O	2	–	–	–	4	2	1

Task

Calculate the following on a daily basis, and then a weekly basis.
- room occupancy
- sleeper occupancy
- double occupancy
- average room rate
- income occupancy

1 hour

6 Aircrews

The hotel where you work takes an aircrew at a reduced rate (12% reduction on apartments). The normal rate is:
SB £35
TB £55

The crew arrives every third day and stays for two nights.
It is usually made up of:

pilot	SB	senior steward/ess	SB
co-pilot	SB	stewards	TB
purser	SB	stewardesses	TB × 2

The airline pays for English breakfast (£3.50) per person and dinner (£7.25) but does not pay for any extras.

1 Calculate the cost to the airline of a normal stay period.
2 From the crew list calculate the total revenue to the hotel for that stay period.
3 Which seven departments would receive a copy of the crew list?
4 What points should be observed when allocating rooms to airline crews?

45 minutes

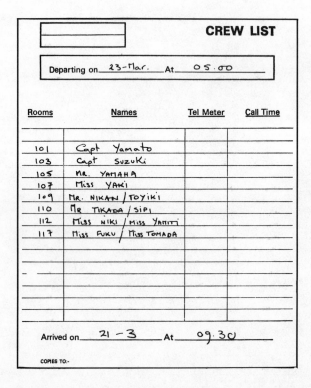

CREW LIST

Departing on __23-Mar.__ At __05.00__

Rooms	Names	Tel Meter	Call Time
101	Capt Yamato		
103	Capt Suzuki		
105	Mr. Yamaha		
107	Miss Yaki		
109	Mr. Nikan / Toyiki		
110	Mr Tikada / Sipi		
112	Miss Niki / Miss Yamiti		
117	Miss Fuku / Miss Tomada		

Arrived on __21 - 3__ At __09.30__

COPIES TO:-

7   Sales statistics

September 1987

	%
Business house	29
Individual	21
Booking agents	9
Travel agents	15
Tour operators	12
Referrals	2.5
Central reservations office	5
Conferences	2
Airline	4.5

Tasks

1 From the information produce a graph or chart suitable for display in the Sales Manager's Office, showing the different sources of reservations in your hotel for the month of September.

2 What suggestions would you give to the Sales Manager regarding the next sales campaign?

45 minutes

8   VAT table

VAT has to be added to all telephone units used when being posted to a guest's account.

Task

Due to an increase of 2p per unit you have been asked to update the quick reference table below. Work out the new prices and the VAT to be charged.

Units	Price	VAT
1	10p	2p
2	20p	3p
3	30p	5p
4	40p	6p
5	50p	8p
6	60p	9p
7	70p	11p
8	80p	12p
9	90p	14p
10	1.00	15p
11	1.10	17p
12	1.20	18p
13	1.30	20p
14	1.40	21p
15	1.50	23p
16	1.60	24p
17	1.70	26p
18	1.80	27p
19	1.90	29p
20	2.00	30p
30	3.00	45p
40	4.00	60p
50	5.00	75p

45 minutes

9 Commissions

Task

1 Work out exactly how much commission a travel agent (Thomas Cook, 100 Park Avenue, New York USA) will receive from the following three invoices if he is being paid 10% commission net. Do not forget to add VAT to the total, at the standard rate of 15%.

2 Enter the information onto the two travel agency commission vouchers.

3 Rule up a cheque in sterling to Thomas Cook and include the following:
Lloyds Bank, High Street, Park Street, London W1
Bank Sorting Code: 20-01-03
Account number: 01063142
Cheque number: 511221
The drawer's name: Walcott Manor Hotel
Sign your own name on the cheque to authorise payment.

4 If you were sending a US $ cheque and the exchange rate was $1.70: £1 how much would Thomas Cook be paid?

Enter in the section marked US $ on the commission advice.

1 hour

WALCOTT MANOR HOTEL
Market Blighton

Travel Agency
Commission Voucher

Guest Name	Arrival	Room	Rate	No. of Nights	% Commission	Total

WALCOTT MANOR HOTEL

Self Billed Commission Invoice No. **04014**

Invoice Date _ _/_ _/_ _

Sub Total	
V.A.T. * (See Note)	
Total $	
Total £	

* Note   The tax shown on this Invoice is the Supplier's output tax.

Agent V.A.T. No.

Hotel V.A.T. No. 242 0289 88

UK 11 C

WALCOTT MANOR HOTEL
Market Blighton

Travel Agency
Commission Voucher

Guest Name	Arrival	Room	Rate	No. of Nights	% Commission	Total

WALCOTT MANOR HOTEL

Self Billed Commission Invoice No. 04014

Invoice Date _ _/_ _/_ _

Sub Total	
V.A.T. * (See Note)	
Total $	
Total £	

* Note   The tax shown on this Invoice is the Supplier's output tax.

Agent V.A.T. No.

Hotel V.A.T. No. 242 0289 88

UK 11 C

COMMISSION ADVICE

NAME OF GUEST	ARRIVAL DATE	RATE	No. NIGHTS	COMMISSION PAYMENT		COMMISSION TO BE DEDUCTED	
				%	£	%	° £

AGENTS VAT No. ....

TOP — TRAVEL AGENT
YELLOW — CASH VOUCHER
BLUE — STATISTICS
PINK — RESERVATIONS

1268

TOTAL	
VAT	
TOTAL	
US $	

TOTAL	
VAT	
TOTAL	
US $	

Authorised by _____
Form A259

1

ROOM No. 115   No. OF GUESTS 1

RATE £45.00 change to £30 on 19/5

NAME Mr. P. James

DATE 18/5-20/5

Δ ····030.40		
····001.00	BAR	
····031.40	TOTL	
····045.00	ROOM CHG	
····000.79	PAID OUT	
000481☆-0	····077.19	
Δ ····077.19		
····050.00	PAID	
····027.19	TOTL	
····030.00	ROOM CHG	
····004.50	TAX	
000482☆-0	····061.69	
Δ ····061.69		
····001.00	BAR	
000483☆-0	····062.69	

Cheques can only be accepted by prior arrangement

CHARGE TO _____   DEPOSIT PAID £ _____

SIGNATURE _____   REFERENCE _____

Please leave your key with the Porter   Bitte lassen Sie Ihren Schlüssel beim Portier

Prière de laisser votre clef à la conciergerie   Sirvanse dejar la llave al conserje

---

2

ROOM No. 105   No. OF GUESTS 2

RATE £45 + vat

NAME Mr. J. Brown

DATE 15th May-18th May

····045.00	ROOM CHG	
····006.75	TAX	
L····004.00	REST	
····001.20	BAR	
000484☆-0	····056.95	
Δ ····056.95		
L····000.80	PHON	
····045.00	ROOM CHG	
····000.75	TAX	
000485☆-0	····103.50	
Δ ····103.50		
····002.50	MISC	
····004.60	BAR	
····045.00	ROOM CHG	
····006.75	TAX	
000486☆-0	····162.35	
Δ ····162.75		
····006.00	TAX	
000487☆-0	····168.75	

Cheques can only be accepted by prior arrangement

CHARGE TO _____   DEPOSIT PAID £ _____

SIGNATURE _____   REFERENCE _____

Please leave your key with the Porter   Bitte lassen Sie Ihren Schlüssel beim Portier

Prière de laisser votre clef à la conciergerie   Sirvanse dejar la llave al conserje

---

3

ROOM No. 119   No. OF GUESTS 2

RATE £45

NAME Mr/Mrs P. Taylor

DATE 15/5-17/5

Δ ····080.20		
····045.00	ROOM CHG	
····000.68	BAR	
····001.02	MISC	
000488☆-0	····126.90	
Δ ····126.90		
····012.69	TAX	
····139.59	TOTL	
····002.00	BAR	
····004.00	ROOM SERV	
····145.59	TOTL	
····045.00	ROOM CHG	
000489☆-0	····190.59	

Cheques can only be accepted by prior arrangement

CHARGE TO _____   DEPOSIT PAID £ _____

SIGNATURE _____   REFERENCE _____

Please leave your key with the Porter   Bitte lassen Sie Ihren Schlüssel beim Portier

Prière de laisser votre clef à la conciergerie   Sirvanse dejar la llave al conserje

10   Tourist statistics

You are working for a tourist office in Pebble Cove which is in a resort area of Great Britain. Your office has recently received figures relating to the number of visitors in your area.

Figures are as follows:

Month	No of visitors	Nationalities	
Jan	7 200	British	70%
Feb	8 500	North American	3.5%
Mar	11 200	European	15%
Apr	11 420	Japanese	5%
May	11 850	South American	2.5%
Jun	12 555	Other	4%
Jul	13 900		
Aug	12 750		
Sep	11 500		
Oct	5 250		
Nov	5 550		
Dec	7 000		

Local hotels have also released the following information:

Hotel	Yearly occupancy
The Royal Hotel	75.50%
Crown Hotel	68.20%
Tralee Hotel	79.52%
Highcliffe Hotel	82.73%
Savoy Court Hotel	64.30%
White Hermitage Hotel	76.55%
The Cloverleaf Hotel	62.94%
The Grosvenor Court	59.25%
Royal Adelphi	61.56%
Whitecliffe Hotel	64.47%
Blue Sea Hotel	55.85%
The Grand	74.29%

Task

Your supervisor has asked you to produce a chart (line graph, bar chart or pie chart) for display in the office. Display the figures in the way you think most suitable.

1 hour

10 Housekeeping, health and safety

1 Health and safety

Task

1 In small groups of 2 or 3 prepare a checklist of *possible* safety hazards in one area of your college, e.g. training restaurant, kitchen, reception office, corridors and staircases.

2 Exchange your list with another group and inspect that area, noting any actual hazards, their likely effect and the action you would recommend to prevent an accident occurring.

1 hour

2 Uniforms

You have been asked to submit designs for uniforms for the staff at a new hotel which will be opening shortly.

Tasks

1 As a group, decide upon a colour which would be acceptable as the house colour for this establishment.
2 In smaller groups, take one department each, e.g. Housekeeping, Reception, Restaurant, Bars, Banqueting, and design a selection of uniforms for the staff in that department, keeping to the house colours.
3 Supply rough drawings of the style of uniform you have designed.
   (a) Give details of the fabric which you recommend, giving reasons for your choice.
   (b) Indicate how you would expect these uniforms to be cleaned and maintained.

1 hour

3   Security

You are the duty housekeeper on an evening shift at the Garden Hotel. You receive a telephone call from a Mr Reynolds who was recently staying at the hotel: he thinks he left a pair of shoes in the bedroom and he is making enquiries as to their whereabouts.

Tasks

1 List the information you would require from Mr Reynolds before you could begin your investigation.
2 List the steps you would take in an effort to locate the shoes.

45 minutes

4 Provision of guest supplies

The Hilbren Hotel is a three star hotel with 150 bedrooms all with en suite facilities.

The management of the hotel are reconsidering the provision of guest supplies and giveaways placed in each room.

Currently guests are provided with:
- one hand soap per person
- one bath soap per room
- one box tissues per room
- two rolls toilet paper per room
- one book of matches per person
- two sheets of hotel notepaper per room
- two envelopes per room

Tasks

1 In groups of three or four:
(a) Prepare a list of additional supplies and giveaways which could be placed in guest rooms.
(b) From this list, choose six items which the group consider to be the most suitable, stating the number suggested for each guest/room and giving reasons for their selection.
2 As a class produce a list of the six most commonly selected items and discuss the reasons why the groups considered these to be the most suitable.

1 hour

5   First Aid

As a Floor Housekeeper at the Firdix Hotel you have recently attended a First Aid course and are now one of the Hotel's registered First Aiders responsible for attending to the needs of both staff and guests.

On Thursday 15th November you are called to the reception area where a guest, Mrs Thompson, has tripped and fallen over a piece of loose carpet. When you arrive at Reception you find Mrs Thompson sitting on a chair looking pale and shocked. She tells you that she has hurt her ankle and when you examine this you find that the ankle is beginning to swell and is causing pain as you touch it.

Tasks

1 Describe the First Aid procedure you would take when confronted with this accident.
2 Prepare a list of information you would need to collect in order to complete an Accident Report.
3 Discuss the legal implications of the accident to both the establishment and the guest.
4 Discuss the methods which could have been used to prevent this accident from occurring.

1 hour

6 Fire

Tasks

1 Prepare a study of the fire policy of one establishment, preferably one which has sleeping accommodation. Use either an establishment where you are or have been on industrial release, or one which you can visit easily.

Give information on:
(a) the staff training programme
(b) fire precautions, e.g. alarms, escape routes, fire doors, extinguishers
(c) procedures in the event of fire
(d) information given to residents, e.g. fire notices

2 When you have completed the study, discuss with the other members of your group the similarities and the differences between each establishment's fire policy.

2 hours

7 Staffing

The White Rose Hotel has 224 rooms with accommodation for 372 guests. There are four floors of bedrooms; each floor is identical in layout with 18 single and 38 twin bedded rooms per floor. Every room has a private bathroom en suite.

Room maids are required to work an eight hour shift with a 15 minute break for coffee and a 30 minute break for lunch. There is an allowance of 35 minutes for packing their trolley, cleaning the service room at the end of their shift and for vacuuming the corridor. They are allowed an average of 25 minutes in which to complete one room.

Tasks

1 Calculate the number of rooms each room maid is required to service in one day.

The Housekeeping Department receives the occupancy forecast for the following week:

	Sunday	Monday	Tuesday	Wednesday	Thursday	Friday	Saturday
Date	6/4	7/4	8/4	9/4	10/4	11/4	12/4
No of rooms	104	198	220	224	224	154	166
% occupancy	46%	88%	98%	100%	100%	69%	74%

2 Calculate the number of room maids required for each of the above days. (Remember – an occupancy forecast refers to the number of occupied rooms per *night*.)

All room maids work a five day week.

3 Calculate the total number of room maids required for the week.

1½ hours

8 Recruitment

The Unicorn Hotel is a city centre hotel with 185 bedrooms, each with a bathroom en suite. The hotel caters mainly for business people and is frequently fully occupied on Tuesday, Wednesday and Thursday nights. Although the management have advertised a series of reduced rate weekend breaks these have only been of moderate success and the hotel is often half-empty at the weekend.

The hotel is experiencing problems in recruiting suitable staff to work as room maids, house porters and linen room assistants. Currently all staff are expected to work a five day, 40 hour week on a rota basis with every third weekend off. The wages offered are similar to those paid by other hotels in the area.

Tasks

1 Discuss the various methods which the hotel could use to attract more staff to the vacant jobs.

2 (a) Study recent trade publications and select two advertisements for each of the jobs.

(b) Compare the two advertisements, noting the job information included in each one.

3 Draft an advertisement for the Unicorn Hotel for one of the jobs, including all the information which you think is important to attract potential employees.

1½ hours

9 Training

You have recently been asked by your departmental manager to take responsibility for training new operative staff.

Tasks

1 (a) Select a piece of electrical equipment, e.g. a suction cleaner and attachments, or a rotary floor machine.
(b) List all the cleaning tasks using the machine which would have to be shown to a new employee.
2 Prepare a training programme which could be used to demonstrate some or all of these cleaning tasks over a period of no longer than five days.
3 Give a talk/demonstration of approximately 10 minutes to a group of two or more people on one of the cleaning tasks you have identified.

2 hours

10 Work planning

Look at the plan of the ground floor public areas of the Floral Hotel.

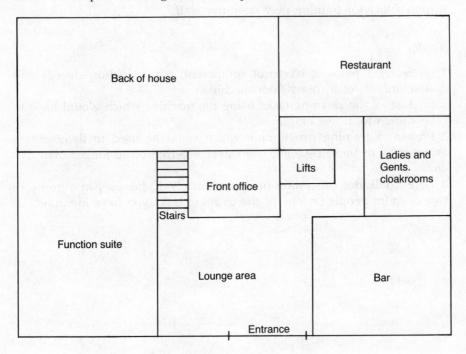

Tasks

Working in groups of 3 or 4:

1 Identify the types of surfaces, furniture and furnishings likely to be found in these areas.

2 Identify the *daily* cleaning tasks which you consider would be required in each area – include any tasks which may need to be carried out more than once per day.

3 Draft out the order of work you would give to the person or persons who would be responsible for cleaning these areas.

2–3 hours (1 hour follow-up)

11 Integrated assignments

1 Advance reservations

The following information relates to advance reservations, and guest arrivals.

Task

The following people are arriving today and should be entered:
   (a) on the density chart,
   (b) in the hotel diary
   (c) when they arrive, onto the bedsheet.

Mr/s Silver	TB	5 nights
Mr/s Rabkin	TB	10 nights
Mr/s Paul	TB	4 nights
Mr/s White	TB	2 nights
Mr/s O'Reilly	TB	9 nights
Miss Wilson	SB	7 nights
Mr Hawkins	SB	8 nights
Mrs Conner	SB	3 nights
Fam Scott	SB	
	TB	10 nights

The next day

Mr/s Guy	TB	1
Mr/s Newman	TB	9
Mr/s Dubois	TB	10
Mr/s Beam	TB	8
Mr/s Wood	TB	6
Mrs Cutter	SB	8
Mr Schmitt	TB	7
Mr McClusky	SB	5
Mr/s Weller	TB	5
Mrs Johnson	SB	4

2 hours

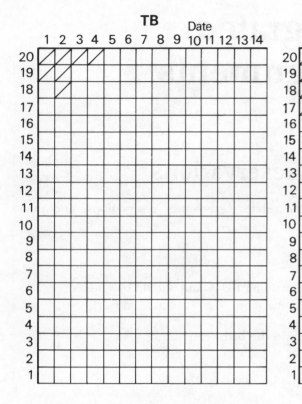

TB Date

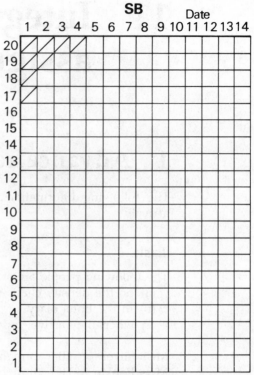

SB Date

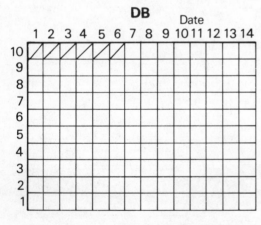

DB Date

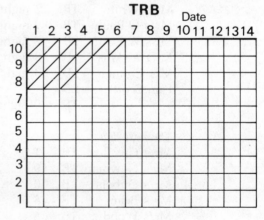

TRB Date

Day 1

NAME	Type of Room	No. of Nights	RATE	How Booked	Confirm-ation	Room No.	REMARKS

Day 2

NAME	Type of Room	No. of Nights	RATE	How Booked	Confirm- ation	Room No.	REMARKS

Day 1

TWIN WITH BATH

ROOM		NAME		STAY	REMARKS	MIX CODE	VIS	£	p.
		2nd. FLOOR HOTEL							
200	E								
201	S	RAY	MR/S	1			2	40	00
209	N								
x211	N								
x215	N								
219Sh.	N	HOWE	MR/S	3	VIP		2	40	00
223	N								
225	N								
229	N								
237	E	McKIE	MR/S	1			2	40	00
238	W								
243	E								
		3rd. FLOOR HOTEL							
301	S								
309	N								
x311	N	DAVIES	MR/S	5			2	40	00
x315Sh.	N								
319Sb.	N								
323	N								
325	N	CURNOW	MR/S	2			2	40	00
337	E								
338	N								
348	W				Comm. with 350				
		4th. FLOOR HOTEL							
400	E								
401	S				Comm.				
407	N								
409	N	BARKER	MR/S	1	C.O.		2	40	00
x411	N								
415	N	AMERICO	MR/S	7			2	40	00
419Sh.	N								
x429	N								
437	E	CARELESS	MRS	4	TasS	(2)		35	00
439	E								
x447	E								
449	E								
		5th. FLOOR HOTEL							
501	S								
502	E	JAMES		2	AMEX		2	—	
505	W	WILSON		2	AMEX		2	—	
507	N	JOHNS		2	AMEX		2	—	
509	N	WELLS		2	AMEX		2	—	
x511	N	TARRANT		2	AMEX		2	—	
515	N								
517Sh.	N	BARRETT		2	AMEX		2	—	
519Sh.	N	BRIAN		2	AMEX		2	—	
521Sh.	N								
525	N	WELLS		2	AMEX		2	—	
x529	N	JONES		2	AMEX		2	—	
x537Sh.	E								
545Sh.	E								
x547	E								

SINGLE WITH BATH

ROOM		NAME		STAY	REMARKS	MIX CODE	VIS	£	p.
		6th. FLOOR HOTEL							
601Sh.	S								
602	E	SMITH		2	AMEX		1	—	
605	W								
607	N	BROWN		2	AMEX		1	—	
609	N								
x611Sh.	N	SIMPSON		2	AMEX		1	—	
612	S								
616 Sh.	S				No Bath				
618	S								
619Sh.	N								
620Sh.	S	JACKSON	MRS	1			1	30	00
621Sh.	N								
622	S	CARTER	Miss	2			1	30	00
625	N								
x629Sh.	N	BAIRD	Miss	2			1	30	00
635	E								
x637	E	CROZIER	MRS	1			1	30	00
643Sh.	E								
645Sh.	E								
		4th. FLOOR WING							
175	N+W								
		5th. FLOOR WING							
195	N+W								

SUITES

ROOM		NAME	STAY	REMARKS	MIX CODE	VIS	£	p.
		2nd. FLOOR HOTEL						
239 TB	E	EL HASSAN		1 Night Let Only			130	00
241 SR	E							
		3rd. FLOOR HOTEL						
331 SR	N							
333 S	N			1 Night Let Only				
335 TB Sh.	E			" " " "				
339 TB	E			1 Night Let Only				
341 SR	E							
343 SR	E			Z Bed.				
x347 TB	E			1 Night Let Only				
349 SR	S							
350 TB	W			Comm. with 348				
351 SR	S							
		4th. FLOOR HOTEL						
431 SR	N							
435 TB Sh.	E			1 Night Let Only				
		5th. FLOOR HOTEL						
539 TB	E			1 Night Let Only				
541 SR	E							
		6th. FLOOR HOTEL						
x639 TB	E			1 Night Let Only				
641 SR	E							

TOURS

ROOM	NAME	STAY	REMARKS	MIX CODE	VIS	£	p.
	AMERICAN EXPRESS	2	No. 3217			320	—

GRAND TOTALS

157

Day 2

ROOM		NAME	STAY	REMARKS	MIS CODE	VIS	C	P.
TWIN WITH BATH								
2nd. FLOOR HOTEL								
200	E							
201	S							
209	N							
x 211	N							
x 215	N							
219 Sh.	N							
223	N							
225	N							
229	N							
237	E							
238	W							
243	E							
3rd. FLOOR HOTEL								
301	S							
309	N							
x 311	N							
x 315 Sh.	N							
319 Sh.	N							
323	N							
325	N							
337	E							
338	N							
348	W			Comm. with 350				
4th. FLOOR HOTEL								
400	E							
401	S			Comm.				
407	N							
409	N							
x 411	N							
415	N							
419 Sh.	N							
x 429	N							
437	E							
439	E							
x 447	E							
449	E							
5th. FLOOR HOTEL								
501	S							
502	E							
505	W							
507	N							
509	N							
x 511	N							
515	N							
517 Sh.	N							
519 Sh.	N							
521 Sh.	N							
525	N							
x 529	N							
x 537 Sh.	E							
545 Sh.	E							
x 547	E							
SINGLE WITH BATH								
6th. FLOOR HOTEL								
601 Sh.	S							
602	E							
605	W							
607	N							
609	N							
x 611 Sh.	N							
612	S							
616 Sh.	S			No Bath				
618	S							
619 Sh.	N							
620 Sh.	S							
621 Sh.	N							
622	S							
625	N							
x 629 Sh.	N							
635	E							
x 637	E							
643 Sh.	E							
645 Sh.	E							
4th. FLOOR WING								
175	N.W							
5th. FLOOR WING								
195	N.W							

ROOM		NAME	STAY	REMARKS	MIS CODE	VIS	C	P/
SUITES								
2nd. FLOOR HOTEL								
239 TB	E			1 Night Let Only				
241 SR	E							
3rd. FLOOR HOTEL								
331 SR	N							
333 S	N			1 Night Let Only				
335 TB Sh.	E			" " " "				
339 TB	E			1 Night Let Only				
341 SR	E							
343 SR	E			Z Bed.				
x 347 TB	E			1 Night Let Only				
349 SR	S							
350 TB	W			Comm. with 348				
351 SR	S							
4th. FLOOR HOTEL								
431 SR	N							
435 TB Sh.	E			1 Night Let Only				
5th. FLOOR HOTEL								
539 TB	E			1 Night Let Only				
541 SR	E							
6th. FLOOR HOTEL								
x 639 TB	E			1 Night Let Only				
641 SR	E							

2 Density chart/room occupancy

Since the entries on the density chart below there have been the following reservations and amendments:

- Niknam Tours request one further night's accommodation. Now departing on the 8th. Accommodation comprised of 3TB, 8DB, and 4SB.
- Mr/s Jones would like two extra SB for the 3rd.
- Baird Incorporated would like the following from the 17th for four nights: 2SB, 6TB, 4DB and 1TRB.
- Mr Astles requests 1TRB from the 13th for six nights.
- Blake Travel would like 1SB and 1TB from the 9th for two nights.

SB	1	2	3	4	5	6	7	8	9	10	11	12	13	14	15	16	17	18	19	20
20	ø	ø	ø	ø	ø	ø	ø	ø	ø	ø	ø	ø	ø	ø	ø	ø	ø	ø	ø	ø
19	ø	ø	ø	ø	ø	ø	ø	ø	ø	ø	ø	ø	ø	ø	ø	ø	ø	ø	ø	ø
18	ø	ø	ø	ø	ø	ø	ø	ø	ø	ø	ø	ø	ø	ø	ø	ø	ø	ø	ø	ø
17	ø	ø	ø	ø	ø	ø	ø	ø	ø	ø	ø	ø	ø	ø	ø	ø	ø	ø	ø	ø
16	ø	ø	ø	ø	ø	ø	ø	ø	ø	ø	ø	ø	ø	ø	ø	ø	ø	ø	ø	ø
15	ø	ø	ø	ø	ø	ø	ø	ø	ø	ø	ø	ø	ø	ø	ø	ø	ø	ø	ø	ø
14	ø	ø	ø	ø	ø	ø	ø	ø	ø	ø	ø	ø	ø	ø	ø	ø	ø	ø	ø	ø
13	ø	ø	ø	ø	ø	ø	ø	ø	ø	ø	ø	ø	ø	ø	ø	ø	ø	ø	ø	ø
12	ø	ø	ø	ø	ø	ø	ø	ø	ø	ø	ø	ø	ø	ø	ø	ø	ø	ø	ø	ø
11	ø	ø	ø	ø	ø	ø	ø	ø	ø	ø	ø	ø	ø	ø	ø	ø	ø	ø	ø	ø
10	ø	ø	ø	ø	ø	ø	ø	ø	ø	ø	ø	ø	ø	ø	ø	ø	ø	ø	ø	ø
9	ø	ø	ø	ø	ø	ø	ø	ø	ø	ø	ø	ø	ø	ø	ø	ø	ø	ø	ø	ø
8	ø	ø	ø	ø	ø	ø	ø	ø	ø	ø	ø	ø	ø	ø	ø	ø	ø	ø	ø	ø
7	ø	ø	ø	ø	ø	ø	ø	ø	0	0	ø	ø	ø	ø	ø	ø	ø	ø	ø	ø
6	0	ø	0	ø	ø	ø	ø	ø	0	0	ø	ø	ø	ø	ø	ø	ø	ø	ø	ø
5	0	ø	0	0	ø	0	0	0	0	0	0	ø	ø	ø	ø	ø	ø	ø	ø	ø
4	0	ø	0	0	0	0	0	0	0	0	0	ø	0	ø	ø	ø	ø	0	ø	ø
3	0	0	0	0	0	0	0	0	0	0	0	0	0	0	0	ø	ø	0	0	ø
2	0	0	0	0	0	0	0	0	0	0	0	0	0	0	0	ø	0	0	0	0
1	0	0	0	0	0	0	0	0	0	0	0	0	0	0	0	0	0	0	0	0

TRB	1	2	3	4	5	6	7	8	9	10	11	12	13	14	15	16	17	18	19	20
10	ø	ø	ø	ø	ø	ø	ø	ø	ø	ø	ø	ø	ø	ø	ø	ø	ø	ø	ø	ø
9	ø	ø	ø	ø	ø	ø	ø	ø	ø	ø	ø	ø	ø	ø	ø	ø	ø	ø	ø	ø
8	ø	ø	ø	ø	ø	ø	ø	ø	ø	ø	ø	ø	ø	ø	ø	ø	ø	ø	ø	ø
7	ø	ø	ø	ø	ø	ø	ø	ø	ø	ø	ø	ø	ø	ø	ø	ø	ø	ø	ø	ø
6	ø	ø	ø	ø	ø	ø	ø	ø	ø	ø	ø	ø	ø	ø	ø	ø	0	ø	ø	ø
5	0	ø	0	0	ø	ø	ø	ø	ø	ø	0	0	0	0	0	0	ø	0	0	0
4	0	0	0	0	0	0	ø	ø	0	0	0	0	0	0	0	0	0	0	0	0
3	0	0	0	0	0	0	0	0	0	0	0	0	0	0	0	0	0	0	0	0
2	0	0	0	0	0	0	0	0	0	0	0	0	0	0	0	0	0	0	0	0
1	0	0	0	0	0	0	0	0	0	0	0	0	0	0	0	0	0	0	0	0

DB	1	2	3	4	5	6	7	8	9	10	11	12	13	14	15	16	17	18	19	20
20	Ø	Ø	Ø	Ø	Ø	Ø	Ø	Ø	Ø	Ø	Ø	Ø	Ø	Ø	Ø	Ø	Ø	Ø	Ø	Ø
19	Ø	Ø	Ø	Ø	Ø	Ø	Ø	Ø	Ø	Ø	Ø	Ø	Ø	Ø	Ø	Ø	Ø	Ø	Ø	Ø
18	Ø	Ø	Ø	Ø	Ø	Ø	Ø	Ø	Ø	Ø	Ø	Ø	Ø	Ø	Ø	Ø	Ø	Ø	Ø	Ø
17	Ø	Ø	Ø	Ø	Ø	Ø	Ø	Ø	Ø	Ø	Ø	Ø	Ø	Ø	Ø	Ø	Ø	Ø	Ø	Ø
16	Ø	Ø	Ø	Ø	Ø	Ø	Ø	Ø	Ø	Ø	Ø	Ø	Ø	Ø	Ø	Ø	Ø	Ø	Ø	Ø
15	Ø	Ø	Ø	Ø	Ø	Ø	Ø	Ø	Ø	Ø	Ø	Ø	Ø	Ø	Ø	Ø	Ø	Ø	Ø	Ø
14	Ø	Ø	Ø	Ø	Ø	Ø	Ø	Ø	Ø	Ø	Ø	Ø	Ø	Ø	Ø	Ø	Ø	Ø	Ø	Ø
13	Ø	Ø	Ø	Ø	Ø	Ø	Ø	Ø	Ø	Ø	Ø	Ø	Ø	Ø	Ø	Ø	Ø	Ø	Ø	Ø
12	Ø	Ø	Ø	Ø	Ø	Ø	Ø	Ø	Ø	Ø	Ø	Ø	Ø	Ø	Ø	Ø	Ø	Ø	Ø	Ø
11	Ø	Ø	Ø	Ø	Ø	Ø	Ø	Ø	Ø	Ø	Ø	Ø	Ø	Ø	Ø	Ø	Ø	Ø	Ø	Ø
10	Ø	Ø	Ø	Ø	Ø	Ø	Ø	Ø	Ø	Ø	Ø	Ø	Ø	Ø	0	Ø	Ø	Ø	Ø	Ø
9	Ø	Ø	Ø	Ø	Ø	Ø	Ø	Ø	Ø	Ø	Ø	Ø	Ø	Ø	0	Ø	0	Ø	Ø	0
8	Ø	Ø	Ø	Ø	Ø	0	Ø	0	Ø	0	Ø	Ø	Ø	0	Ø	0	Ø	0	Ø	0
7	Ø	Ø	Ø	Ø	0	0	0	0	0	0	Ø	Ø	Ø	0	0	0	Ø	0	0	0
6	0	Ø	Ø	Ø	Ø	0	0	0	0	0	0	Ø	0	0	0	0	Ø	0	0	0
5	0	Ø	Ø	Ø	Ø	0	0	0	0	0	0	Ø	0	0	0	0	0	0	0	0
4	0	Ø	Ø	Ø	0	0	0	0	0	0	0	0	0	0	0	0	0	0	0	0
3	0	Ø	Ø	Ø	0	0	0	0	0	0	0	0	0	0	0	0	0	0	0	0
2	0	0	Ø	Ø	0	0	0	0	0	0	0	0	0	0	0	0	0	0	0	0
1	0	0	0	0	0	0	0	0	0	0	0	0	0	0	0	0	0	0	0	0

TB	1	2	3	4	5	6	7	8	9	10	11	12	13	14	15	16	17	18	19	20
20	Ø	Ø	Ø	Ø	Ø	Ø	Ø	Ø	Ø	Ø	Ø	Ø	Ø	Ø	Ø	Ø	Ø	Ø	Ø	Ø
19	Ø	Ø	Ø	Ø	Ø	Ø	Ø	Ø	Ø	Ø	Ø	Ø	Ø	Ø	Ø	Ø	Ø	Ø	Ø	Ø
18	Ø	Ø	Ø	Ø	Ø	Ø	Ø	Ø	Ø	Ø	Ø	Ø	Ø	Ø	Ø	Ø	Ø	Ø	Ø	Ø
17	Ø	Ø	Ø	Ø	Ø	Ø	Ø	Ø	Ø	Ø	Ø	Ø	Ø	Ø	Ø	Ø	Ø	Ø	Ø	Ø
16	Ø	Ø	Ø	Ø	Ø	Ø	Ø	Ø	Ø	Ø	Ø	Ø	Ø	Ø	Ø	Ø	Ø	Ø	Ø	Ø
15	Ø	Ø	Ø	Ø	Ø	Ø	Ø	Ø	Ø	Ø	Ø	Ø	Ø	Ø	Ø	Ø	Ø	Ø	Ø	Ø
14	Ø	Ø	Ø	Ø	Ø	Ø	Ø	Ø	Ø	Ø	Ø	Ø	Ø	Ø	Ø	Ø	Ø	Ø	Ø	Ø
13	Ø	Ø	Ø	Ø	Ø	Ø	Ø	Ø	Ø	Ø	Ø	Ø	Ø	Ø	Ø	Ø	Ø	Ø	Ø	Ø
12	Ø	Ø	Ø	Ø	Ø	Ø	Ø	Ø	Ø	Ø	Ø	Ø	Ø	Ø	Ø	Ø	Ø	Ø	Ø	Ø
11	Ø	Ø	Ø	Ø	Ø	Ø	Ø	Ø	Ø	Ø	Ø	Ø	Ø	Ø	Ø	Ø	Ø	Ø	Ø	Ø
10	0	Ø	Ø	Ø	Ø	Ø	Ø	Ø	Ø	Ø	Ø	Ø	Ø	Ø	Ø	Ø	Ø	Ø	0	Ø
9	0	Ø	Ø	Ø	Ø	Ø	Ø	Ø	Ø	Ø	Ø	Ø	Ø	Ø	Ø	Ø	Ø	Ø	0	0
8	0	0	Ø	Ø	Ø	Ø	Ø	Ø	Ø	Ø	Ø	Ø	Ø	Ø	Ø	Ø	0	Ø	0	0
7	0	0	0	Ø	Ø	Ø	Ø	Ø	Ø	Ø	Ø	Ø	Ø	0	0	0	Ø	0	0	0
6	0	0	0	Ø	Ø	Ø	Ø	Ø	0	Ø	Ø	Ø	0	0	0	0	0	0	0	0
5	0	0	0	0	0	0	Ø	Ø	0	Ø	Ø	Ø	0	0	0	0	0	0	0	0
4	0	0	0	0	0	0	Ø	0	0	0	0	0	0	0	0	0	0	0	0	0
3	0	0	0	0	0	0	0	0	0	0	0	0	0	0	0	0	0	0	0	0
2	0	0	0	0	0	0	0	0	0	0	0	0	0	0	0	0	0	0	0	0
1	0	0	0	0	0	0	0	0	0	0	0	0	0	0	0	0	0	0	0	0

Tariff

TB	£50	SB	£30
DB	£50	TRP	£70

Task

1 Using this information and the density chart, calculate the following from the 6th to the 12th inclusive on a daily and weekly basis;

    (a) room occupancy

    (b) sleeper occupancy

    (c) average room rate

    (d) income occupancy.

2 hours

3 Systems and organisation

The Russell Hotel will shortly be re-opening after a period of closure for renovation.

- Situation – large provincial town
- Capacity – 120 Rooms, 160 sleepers
- Business mix – business and tourist
- Projected occupancy – 80%
- Length of stay – 70% one night, 25% 2 nights, 5% longer
- Ratio booked to chance – 60% to 40%

Visitors' accounts are dealt with in a separate bill office: the hall porter deals with most enquiries, and therefore you need not concern yourself with either. All guests' confirmations are typed by the Manager's secretary, although all correspondence is filed in the reception office.

Tasks

1 Describe in detail the reception/advance booking system which you would adopt, giving reasons for your choice.

2 Produce an organisation chart for the front office area staffing.

You are advised to concentrate on the actual system employed, showing examples where helpful, or using diagrams to aid clarification.

4 Research

Tasks

1 Working in groups of 3–4 visit at least one hotel and one other recreational facility.

2 Identify the market factors which have influenced the provision of the services which you survey.

3 Compile a questionnaire about the provision of hotels and recreational facilities in your area and interview a representative selection of

(a) providers

(b) customers

4 From the results of your questionnaire, prepare a brief presentation for the rest of the group, outlining the results of your findings.

Note:

- choose somewhere near enough to college or your home so you can make more than one visit if necessary.
- within your entire group, try to cover a reasonably large area.
- decide upon the information you will have to collect.
- decide upon a deadline for assembling the information.
- make a list of the type of things you will have to look for to answer questions relating to the market factors.
- make a list of the things you will put in your questionnaire – be sure the results will actually tell you something.
- decide how many people you aim to question and which age groups – be realistic and don't try to do too many.
- check the local papers for advertisements and see if what is actually provided ties up in any way with what people say they want.

5 Cash control

Task

1 Describe the major methods of payment in use in hotels, and evaluate each one from the point of view of security and liquidity for both the client and the establishment.

Your assignment should be illustrated with:

- tables
- bar charts
- pie charts
- line graphs
- illustrations, where applicable

2 Assess the relative value of each method of payment to the hotel.

6 Organise a party

Your group has decided to arrange and organise a Childrens' Christmas party.

You intend to invite children from the local orphanage (25 in total) and will also invite children of the college staff and friends, at a small fee to cover costs.

Tasks

1 Nominate a committee and allocate responsibilities as appropriate.

2 Compile a questionnaire to circulate to staff so that you can find out favourable dates, ages of children, etc.

3 Write letters to well-known companies asking for samples of their products.

4 Design a poster to place around the college.

5 Design a ticket and check the costs of printing.

6 Compile a programme of events.

7 Arrange a meeting with your tutor and present a report of events to date from each section of the committee.

Note

This project can be expanded, and can be very successfully carried out by an enthusiastic group of students (and staff).

For those taking 720 it is an invaluable assessment of your ability to work together, and for those taking BTEC it is a good test of your organisational ability.

1 term

7 Cash books

The childrens' party which you have arranged is well advanced in preparation but funds are rather short.

You decide to hold a raffle, and ask all the students and staff to bring in items for a food hamper. Once the hamper is arranged it looks very attractive, and the raffle tickets sell well. Money is also coming in from ticket sales, and you need to keep some records.

Tasks

1 Arrange the income and expenditure for the month in the form of a simple cash book

1 Nov	Party ticket sales	2.50
2	Cost of book of raffle tickets	4.80
3	Party sales	5.00
4	Party sales	15.00
4	Purchase of hamper	3.50
5	Raffle sales	2.20
5	Party sales	25.00
8	Raffle sales	3.80
8	Raffle sales	4.40
9	Payment for printing	14.80
13	Raffle sales	5.80
16	Party sales	7.50
16	Raffle sales	4.20
20	Postage costs	2.25
20	Party sales	2.50
21	Raffle sales	4.90
21	Bought balloons etc.	10.75
23	Raffle sales	2.20
27	Party sales	27.50
26	Bought gifts for party – 20 stockings @ 98p each plus VAT	
28	Raffle sales	1.80
30	Party sales	12.50

2 If the party tickets cost £2.50 each how many children are paying to come so far?

3 How much money is in the kitty at the end of November?

30 minutes

Appendix

Hotel A

540 rooms all with private facilities
5 star
6 floors
No grounds (city centre)
Car park
Leisure centre and pool
1 restaurant
1 coffee shop (24 hours)
2 bars
1 wine bar (used extensively by public)
Banqueting room
4 interlinking conference rooms
24 hour room and lounge service

Clientele
Mainly business, some tourists –
discourages package tours although will
occasionally accept them. Regular air crew.

Hotel C

30 rooms – 10 with private bathrooms
1 restaurant (busy)
1 large function room
1 bar
No grounds (outskirts of provincial town)
Car park
Bar lunches/dinners
No room service – courtesy trays
Vending machine in bar area
2/3 star

Clientele
Mainly business with local trade in bar/
restaurant.

Hotel B

167 Rooms: 90 in new wing with private
bathrooms; the old wing has
bathrooms on the floors
1 restaurant
2 bars (both open to public)
3 star
1 small banqueting/conference room
Room service for breakfasts only
24 hour lounge service
No grounds (town centre)
No car park

Clientele
Mixture of business/tourist – often
very full with package tours during
season.